TOWARDS SUSTAINABILITY

Julianne (Jackie) Venning specialises in environment and natural resource management. Her earlier work involved the conservation of rural landscapes, particularly the retention, rehabilitation and re-establishment of remnant native vegetation. More recently she was responsible for the South Australian State of the Environment Reporting program. In this role she was involved in the development of environmental reporting systems at both a state and national level. She holds a PhD in biological sciences and an MBA.

John Higgins worked on the Commonwealth State of the Environment Reporting program for three years. He holds a PhD in chemistry and an MA in science, technology and society. He was a postdoctoral fellow at Oxford and the Australian National University before joining the Australian Public Service.

To
Sam and Charley

To
their Future

TOWARDS SUSTAINABILITY

EMERGING SYSTEMS FOR INFORMING SUSTAINABLE DEVELOPMENT

edited by
Jackie Venning and John Higgins

UNSW PRESS

The views expressed by the contributors in this work are not necessarily those of the editors or their employers.

A UNSW Press book

Published by
University of New South Wales Press Ltd
University of New South Wales
UNSW Sydney NSW 2052
AUSTRALIA
www.unswpress.com.au

© UNSW Press 2001
First published 2001

National Library of Australia
Cataloguing-in-Publication entry:

    Towards sustainability: emerging systems for informing sustainable development.

    Bibliography.
    Includes index.
    ISBN 0 86840 667 8.

    1. Sustainable development — Australia. 2. Sustainable development — Social aspects — Australia. I. Venning, J. (Jackie). II. Higgins, John, 1962– .

    338.994

Printer Griffin Press, Adelaide

CONTENTS

ACKNOWLEDGMENTS

The editors would like to acknowledge the valuable assistance of the following people during the preparation of this book: Dr Tom Beer, CSIRO Atmospheric Research; Dr Geoffrey Bishop, Bishop and Associates, Basket Range, South Australia; Mark Faulkner, South Australian Department for Water Resources; Allan Haines, Environment Australia (now retired); Associate Professor Ronnie Harding, Centre for Environmental Studies, University of New South Wales; Dr Steve Hatfield-Dodds, Environment Australia; Dr Andrew Lothian, South Australian Department for Environment and Heritage; Brett Odgers, Environment Australia (now retired); Dr Ray Wallis, Western Australian Department of Environmental Protection; Dr David Williams, CSIRO Energy Technology; Professor David Yencken, Landscape, Architecture and Environmental Planning, University of Melbourne.

FOREWORD

We cannot work to create a future which we do not first imagine.

If humanity is to live on this beautiful planet of ours indefinitely, we must design and innovate within the next generation or so, the means of creating a sustainable society, and we must complete a mission to realise it. This in turn requires that we envision, design and create sustainable prosperity, which involves the simultaneous advancement of four forms of prosperity: economic, ecological, social and cultural.

To create a sustainable society we must be able to imagine it, model it and understand how it would behave. We cannot work to create a future that we do not first imagine, and we must be able also to imagine how it would work through the creation of appropriate metaphors and models. We must also be able to measure our progress towards the realisation of a sustainable society through the creation of appropriate indicators and assessment processes. Finally we must innovate many new products, services and technologies to market to the world's peoples to provide the tools that would enable them to make this heroic transformation on the ground.

The innovations we will need to create should not only make us less unsustainable but more sustainable as well. To treat illness and to cease to be sick is very different to creating health. To lessen a bad outcome is not the same as creating a good outcome. To improve the efficiency and effectiveness of fossil fuel engines is not the same as creating a hydrogen powered alternative and to reduce waste is not the same as abolishing it.

We need also to create many new ways to measure change in our world. For example, we need a true partnership of economics and ecology in the provision of indicators if we are to assess whether we are achieving economic and ecological prosperity, as win–win rather than win–lose. We must be able to measure whether we are doing economically well while and by doing ecological good, or whether we are still continuing the ways of the modernist past by doing economically well by doing ecological bad. Our economists must incorporate the value of natural capital and of the environmental services provided by nature, such as the production of clean water or the pollinating roles of insects and birds, into our measures of prosperity. This will achieve real progress in our capabilities to measure prosperity in all its forms, including in concepts such as the 'triple bottom line', or as I would prefer, 'the quadruple bottom line'.

To realise a sustainable society and to design and innovate our way to sustainability, we need to create what I call green ways and green wares. Green ways are the values and attitudinal shifts, the customer preferences, and the professional practices and ethical behaviours we need to progress towards a sustainable future. Green wares are the new designs, products, services and technologies we need to innovate. We need both, and those who innovate and market these will be the new leaders in the new green economy of the twenty-first century. These green ways and green wares will assist us, amongst other things, to live within perpetual solar income, abolish the concept of waste, protect biodiversity and avoid or ameliorate all forms of collateral damage. These concepts are among the design rules we need for the creation of a sustainable society.

This book will help our understanding of some core aspects of sustainability, in particular our understanding of the working of both sustainable and unsustainable systems, and it will help us develop the tools to assess whether we are progressing in our journey towards the realisation of a sustainable society. Our hearts tell us that we need a sustainably prosperous society. We however also need to put our best minds and our highest intelligences to the task of building pathways towards its realisation. This book is a significant contribution to this mind work.

Peter Ellyard

Dr Peter Ellyard is currently Executive Chairman of Preferred Futures, and Chairman of the Universal Greening Group of companies and of the MyFuture Foundation. He has been a Senior Adviser to the United Nations system for more than 25 years, including at the 1993 Earth Summit. Peter is the author of the best selling book *Ideas for the New Millennium* (1998, 2001).

LIST OF CONTRIBUTORS

DAVID BENNETT was the Executive Director of the Australian Academy of the Humanities and the Executive Director of National Academies Forum until the end of 2000. During his diverse career, he has also worked on environmental issues for the Aboriginal and Torres Strait Islander Commission, been a policy officer with the Department of the Environment, Sport and Territories and was the Acting Director of the Mawson Graduate Centre for Environmental Studies at the University of Adelaide. He has published on environmental philosophy, particularly environmental ethics, the ethical treatment of non-human animal species, and population issues relating to the environment. David is co-author with Richard Sylvan of *The Greening of Ethics*.

STEPHEN (STEVE) DOVERS is a Fellow with the Centre for Resource and Environmental Studies at the Australian National University. He has spent the past decade researching, writing and communicating in the area of sustainable development, institutional and policy arrangements for resources and environment, and environmental history.

BARNEY FORAN is an ecologist by training and has spent most of his career in the rangelands of Australia, southern Africa and New Zealand. His current task brought him to the Resource Futures group at CSIRO Sustainable Ecosystems in Canberra in 1993. His current research interests focus on radical options for the redesign of Australia's economy that take it to low levels of energy and material usage by 2050. The group is tasked with designing and testing transition pathways 'towards sustainability' for national and regional scales in

Australia. Central to the job has been the development of a number of analytical frameworks that describe how Australia actually works in physical terms ... the so-called 'physical economy'.

ANN HAMBLIN specialises in agricultural land-uses and ecosystem management and currently works at the Bureau of Rural Sciences in Canberra. Ann has worked on international and national land indicators for the World Bank and Environment Australia, and on national indicators of sustainable agriculture for SCARM. She serves on several national and international boards related to environment and agricultural research. She was director of the Cooperative Research Centre for Soil and Land Management 1994–98. In 1999 she convened the Fenner Conference on the Environment called *Visions of Future Landscapes* and more recently she has prepared the land theme for the Australian 2001 State of the Environment report.

JOHN HATCH is a senior lecturer in the School of Economics at Adelaide University. He teaches in several courses in environmental economics and has a special interest in wildlife resources. He has published in the areas of commercialisation of wildlife, waste management and recycling. Most recently he made a major submission to the Senate Rural and Regional Affairs and Transport References Committee on the *Commercial Utilisation of Australian Native Wildlife*. In his spare time he is an avid ornithologist.

TOR HUNDLOE is Professor of Environmental Management at the University of Queensland. He is Chair of the Wet Tropics Management Authority, responsible for the oversight of this World Heritage Area. He also chairs the Nature and Ecotourism Accreditation Program that is responsible for certifying ecotourism products. Tor served as Chair of the Australia Institute in 1999 and 2000 and had been a board member previously. He was Founding President of the Environment Institute of Australia. He trained in economics, political science and environmental management.

FRANZI POLDY is a physicist by training with wide interests in operations research, energy futures and the relationship between the economic and the physical worldviews presented by this analytical approach. His is currently with the Resource Futures group at CSIRO Sustainable Ecosystems in Canberra. His research interests focus on radical options for the redesign of Australia's economy that take it to low levels of energy and material usage by 2050.

1
INTRODUCTION
JOHN HIGGINS AND JACKIE VENNING

'Everyone's an environmentalist now.' So said Australian Prime Minister John Howard at a photo opportunity during the 1998 Federal election campaign. This statement illustrates how environmental issues have recently been absorbed into the political mainstream. Increasing political cognisance of the environment has been accompanied by the growing awareness of sustainable development: most famously defined as 'development that meets the needs of the present, without compromising the ability of future generations to meet their own needs' (WCED 1987).

It is not the intention of this book to enter the debate over particular policies or to argue over the degree to which sustainable development is practised. Rather, proceeding from the premise that sustainable development is the basis of a new, emerging paradigm for corporate, local, national and global activity, the book explores what sort of information will be useful for making decisions in the new milieu. Our interest includes how this information will be generated, what tools and methods will be used, what institutional structures are appropriate to them — and how information will be presented to, and used by, decision makers. In short, this book is about how to inform sustainable development.

DRIVERS OF CHANGE

In the pre-industrial era, the resources and capacity of the world's ecosystems were so large relative to the scale of human activity as to be effectively infinite. There are well-documented cases of pre-industrial civilisations recklessly exploiting, and thus depleting to the point of collapse, the capacity of the environment — in particular, isolated regions. However, human activity was not on a scale to disrupt global ecosystems (Ponting 1991, WCED 1987).

Industrialisation and the rapid growth of the human population have changed the equation. There are many more people, and on average each person makes much greater demands on the environment. People in the developed world are mostly responsible for raising the average demand on global resources. Improving the living standards of the two thirds of the world's population who live in the developing world without imposing intolerable strains on Earth's ecosystems is the problem for which sustainable development has been suggested as a solution (WCED 1987).

Human activity now has the potential to affect the functioning of ecosystems on a global scale. Further, there is strong evidence that it has done so. This evidence is seen most clearly in the depletion of the stratospheric ozone layer (WMO 1995) and in global climate change (IPCC 1995). The environmental effects of other changes, such as massive deforestation, are still being debated but are incontestably adverse (UNEP 1999). The local and regional effects of problems such as pollution and land degradation have also been felt around the globe. In Australia, salinisation and loss of biodiversity associated with land clearing are among the most important environmental challenges.

Awareness of these adverse environmental effects, together with a realisation that the world's population continues to grow and that each person is making, on average, greater demands on the environment, dictates that attitudes to decision making must change.

Simply put, the world must pay more attention to the environment or face catastrophe. The alternatives to continued rapacity are sustainable development or de-industrialisation and a return to a simpler way of life. While some 'deep green' activists urge the latter course, it is unlikely to be acceptable to the vast majority of the world's people, leaving sustainable development as the only prudent and feasible alternative.

HISTORICAL BACKGROUND[1]
INTERNATIONAL DEVELOPMENTS

Our confidence in sustainable development as an emerging paradigm is based largely on the clear historical trend toward greater awareness of, and emphasis on, environmental issues, and the concomitant recognition of the need to take environmental as well as economic and social factors into account when planning development.

Environmental problems are not an exclusively modern phenomenon, nor is concern about the environment a recent development. The city of Florence made laws regulating pollution of the Rivers Arno, Sieve and Serchio as long ago as 1477. The roots of modern environmentalism may be found in the nineteenth century when the first environmental non-government organisations (NGOs) were founded, and recognisable conservation movements emerged. National parks and reserve systems were established, and there were efforts to conserve resources and regulate trade in wildlife.

There was something quite new, however, about the nature and intensity of concern about the environment in the late twentieth century. There was a surge in environmentalism in the 1960s and 1970s. Some statistics illustrate how dramatic the changes have been (McCormick 1992):

- in 1992 there were more than 20 000 environmental NGOs worldwide, one third of which were founded after 1972;

- in 1971 only 12 countries had national environment agencies and today few countries lack one;

- as of 1992 there were 250 international environmental treaties in force, three quarters of them established in the last 30 years; and

- the first Green Party was formed in New Zealand in 1972: two decades later there were more than 20 Green political parties around the world, 11 of which were represented in parliaments.

Five factors have been implicated in the explosion in environmental concern of the 1960s and 1970s. Firstly, there were broad social changes that predisposed people in the West, particularly younger people, to question established values and ways of life. Economic prosperity offered the security to question and protest without too much anxiety about the financial future, and at the same time raised questions about whether the drive for material abundance that had absorbed previous generations was really worthwhile. The protest movements against the Vietnam War and patriarchal social structures, and the fight for civil rights created an atmosphere of radicalism.

Secondly, there was a series of environmental disasters during the 1960s that received wide publicity. Two of the most famous were the spill of 117 000 tonnes of crude oil from the *Torrey Canyon* off the west coast of England in March 1967 and the terrible mercury-induced neurological damage suffered by thousands of Japanese in the towns of Minamata and Niigata. Mercury entered the human food supply through seafood contaminated by the wastes that factories discharged into local waters, and the companies were finally forced to admit liability and pay compensation in 1971 (Niigata) and 1973 (Minamata).

Thirdly, there was widespread alarm over nuclear weapons testing. The policy of Mutually Assured Destruction (MAD) alarmed many people, and concern about the effects of fallout from above-ground testing drifting over populated areas added an extra dimension. The environmental and anti-nuclear movements have been, and remain, closely linked.

Fourthly, better scientific understanding of the consequences of human activity, especially since industrialisation, was demonstrating the dangers to the environment and the alarming implications these might have for people. The health effects of pollution were increasingly recognised.

The fifth factor was the publication in 1962 of Rachel Carson's *Silent Spring*. Despite strong opposition from the political, business, and even scientific establishment, Carson's book raised public awareness of the environmental damage being done by pesticides. It was an impassioned, polemic work grounded in solid science. Had it appeared a few years earlier, *Silent Spring* may have had little impact. Appearing when it did, in an era where the counter culture was developing and environmental disasters were widely publicised, the book was an important catalyst for change. Other influential best-selling books depicting an environmental crisis followed, including *The Population Bomb* by Paul Ehrlich (1968) and the Club of Rome's (1972) *Limits to Growth*.

There have been three significant international developments in the debate over the environment since the emergence of modern environmentalism. These are: the Stockholm Conference of 1972, the report of the World Commission on Environment and Development (the Brundtland Report) in 1987, and the Rio Earth Summit in 1992.

The Stockholm Conference was the first high profile international meeting to deal exclusively with environmental issues, and it attracted enormous public interest. Amongst its important features were the debate between 'first' and 'third' world nations on how environment and development should be approached, the involvement of a range of NGOs, and the establishment of the United Nations Environment Programme. The Conference also set the international environmental agenda for years to come by formulating a Declaration, a set of Principles, and an Action Plan.

The Brundtland Report (WCED 1987) was a landmark document, and remains the most important manifestation of the trend toward a more holistic approach to the environment and development that evolved in the 1980s. The tendency of the early environmental movement was to view development and environmental protection as mutually exclusive. Further reflection suggested that ecologically sustainable development trajectories could be found. The environment and development lobbies began to move away from adversarial approaches to environmental issues and towards co-operative solutions; from win–lose scenarios to win–win outcomes.

A critical insight of the Brundtland Report was that social as well as environmental and economic factors must be taken into account as part of a holistic approach in order to reconcile the environment and development properly. This insight was not original; the realisation had been growing throughout the 1980s. The Brandt Report and the World Conservation Strategy were important steps along this path. The Brundtland Report was the apotheosis of this way of thinking, which was encapsulated by the Brundtland definition of sustainable development as 'development that meets the needs of the present, without compromising the ability of future generations to meet their own needs'.

The Earth Summit, held in Rio de Janeiro in 1992, saw more than 100 nations formally commit themselves to sustainable development. The summit produced Agenda 21, an ambitious, relatively detailed plan to achieve sustainable development in the 21st century (UNCED 1992). The summit also expanded the international environmental agenda established in Stockholm. The new prominence given to bio-logical diversity through the Convention on Biological Diversity and a commitment to take action on the enhanced greenhouse effect, expressed through the Convention on Climate Change, were the most important additions. A follow-up meeting, known as Earth Summit + 5, was held in New York in June 1997. Some progress was reported, although many found the achievements disappointing compared with the high aspirations of Rio.

Each of these major international developments has clarified the relationship between the environment and development, and strength-ened commitment to sustainable development. This process has been mirrored on national and local scales. Many nations, and local govern-ments, now have strategies for sustainable development, and the signatories to Agenda 21 report regularly on their progress. Many companies have incorporated sustainable development into their plan-ning (Elkington et al. 1998).

AUSTRALIAN DEVELOPMENTS

In Australia, the first major environmental statutes were enacted in the 1970s, beginning with the New South Wales *State Pollution Control Commission Act 1970* and the Victorian *Environment Protection Act 1970*. The first Commonwealth legislation was the *Environment Protection (Impact of Proposals) Act 1974*. State Governments created agencies to tackle the problems of pollution, and a Federal Environment Department was established in 1971. In keeping with the spirit of the 1970s, these early efforts were aimed at protecting the environment and fixing the problems caused by development.

In the 1970s and early 1980s there were emotive debates over a range of environmental issues. Those with the highest profiles were uranium mining, the flooding of Lake Peddar, sand mining at Fraser Island, and the proposed construction of the Gordon below Franklin Dam. Building and construction unions became involved through the 'green bans', in which labour was withdrawn from projects that were thought to be environmentally damaging. Uranium mining went ahead, and in 1972 Lake Pedder was flooded. However, Commonwealth Government action prevented sand mining at Fraser Island and the construction of a dam on the Gordon River below its junction with the Franklin River.

In these debates, business and conservation interests often adopted strongly adversarial and opposing roles. By 1992 it was possible to develop a *National Strategy for Ecologically Sustainable Development* (Commonwealth of Australia 1992), which was endorsed by all governments and has the broad support of most major players in industry, and the environment movement. Disagreements over the interpretation and application of the principles contained in the strategy remain, but the various interests have a great deal more common ground than they did in the 1970s.

The Regional Forests Agreement (RFA) process is an example of the new more integrated approach. Forests were a commercially valuable resource to the timber industry, many workers and communities relied upon them, indigenous and other people attached important heritage values to them, and conservationists pointed to their important ecological roles, including as habitat for native flora and fauna, as greenhouse sinks, and as systems for purifying water and controlling hydrology. There was intense conflict over the use of forests, which included large scale demonstrations against forestry practices and activities.

In 1992, the *National Forests Policy Statement* (Commonwealth of Australia 1992) set out principles for forest management that took economic, social and environmental concerns into account. Flowing from the statement, the RFA process involved a massive commitment of resources to develop detailed plans for managing forests sustainably, based on an agreed set of principles and objectives and the best available information. Although debate continues over whether individual agreements meet the stated objectives, the process was grounded in sustainable development principles.

Elsewhere, conservation groups are working with industry to establish and implement corporate standards (for example, ISO 14000), codes of practice for key industries such as fishing, forestry and mining, and performance measures to demonstrate the effectiveness of environmental management.

While it is premature to say that sustainable development has become the dominant paradigm, there is a clear historical trend to suggest that it is emerging as such.

INTERESTS, INFORMATION AND DECISIONS

Decisions are not always made rationally, or on the basis of information alone. Interests, history, tradition, politics, prejudice and personality all have an important role. Interests are, perhaps, the most potent factor of all. Where there is broad agreement on desirable outcomes, or interests coincide, decisions are simplified, although there may still be disagreement over the best course of action. Where differing interests are involved, some way of balancing them must be found. Sometimes, one interest predominates while others are ignored. Such solutions are rarely stable in the long term, especially in democratic societies.

Equally, it is overly cynical to suggest that information plays no role in decision making. On the contrary, the role of information in decision making is often critical. Information allows better judgements about how various interests can be advanced, and may change perceptions of what is in the best interests of various stakeholders. Where there is an agreed framework for balancing interests, the information needs are likely to be most clearly articulated. For example, the Reserve Bank of Australia has clear policies for deciding whether to change official interest rates, and a clearly articulated set of indicators (consumer price index, average wages, the current account, gross domestic product) that are relevant to that decision.

Interests and information are closely related. As a rule, the information that a society, government, corporation, or other entity generates is required in order to advance its interests. This does not mean that information is simply a species of propaganda; information may advance interests by enabling balanced and informed decisions about critical issues as well as being used to sway political arguments.

The emergence of sustainability as a paradigm for development involves a significant shift in the perceived interests of individuals, companies, societies and governments. The type of information required to achieve those interests must also change. In line with these changes, both government and industry tend increasingly to generate their own environmental information, rather than being informed mainly through universities, research institutions and environmental lobby groups.

The development of state of the environment reporting is an example. The first state of the environment report in Australia was little more than a compendium of available environmental facts with minimal commentary (Commonwealth of Australia 1986). It was followed in 1996 by *Australia: State of the Environment 1996* (SEAC 1996), the first comprehensive scientific assessment of the nation's environment. The Australian States and Territories also produce state of the environment reports, the first comprehensive report being *The*

State of the Environment Report for South Australia (Environment Protection Council 1988).

Corporate environmental reporting is following a similar trajectory. The Earth Summit called for regular reporting by corporations on the environmental aspects of their operations. Early reports often took the form of glossy brochures promoting 'good news stories', and some corporate environmental reports still have this character. Increasingly, though, corporations are producing balanced, objective assessments of their progress. Leading corporate environmental reports include quantitative performance measures, full disclosure of breaches of environmental regulations, and assessments of progress against realistic targets and benchmarks.

Bodies such as the World Industry Council for the Environment and the World Bank have published guidelines and indicators for corporate environmental reporting. As of mid-1998, some 600 corporations had produced corporate environmental reports, and some companies are also reporting on social aspects of their operations in corporate sustainability reports.

In Australia, the *Company Law Review Act 1998* amended the Corporations Law to require limited corporate environmental reporting. Directors must report on compliance with environmental regulations, although small proprietary companies are exempt. Additional reporting is at the discretion of the company, and Australian practices generally lag behind those in Europe and North America (Fayers 1997, Deegan 1998). There are some promising signs, however (Elkington 1999). Major resource companies have led the way, and some are near to best practice. Western Mining Corporation released a relatively comprehensive corporate environmental report in 1995, and BHP a similarly comprehensive document in 1997. Both companies have since released annual environmental reports.

THE NEED FOR NEW SYSTEMS

In this book, it is argued that new systems are required to inform decision makers in the context of sustainability. Specifically, it is suggested that the emerging system must be broadly based, emphasise economic, environmental and social considerations equally, use suitable tools and models, and be expressed through institutions that will both generate the information and respond to the 'signals' received through such information. Three points about the evolution of new systems follow from the preceding discussion.

Firstly, the need to develop new systems was not identified until recently. This is partly because the concept of sustainability was articulated only in the last 20 years (although its intellectual roots twine through many centuries — see Chapter 2), but also because humans have only recently recognised the importance of environmental considerations in development. Secondly, a time lag is to be expected between the emergence of the concept of sustainability and the development of systems appropriate to a sustainable society. Thirdly, the drivers behind changes are very powerful. However, it does not follow from this that systems will swing completely to the ideal forms described in this book. In practice there are a host of interlinked systems that will vary considerably in how nearly they approach the models described here.

INFORMING SYSTEMS

This book deals with the whole gamut of processes and apparatus involved in selecting data, processing it to generate information, communicating that information to those who need it, and using it to make decisions. The boundaries of the system are thus broad, encompassing the users as well as the generators of information. We use the term 'informing systems' (Dovers 1996) to denote this complex mix. The expression is deliberately wider than 'information system', because our brief is broader, extending well beyond the question of how data are processed to create information.

Informing systems are not necessarily formally organised, self-contained, or even readily identified. One person or organisation may be part of several overlapping informing systems concerned with delivering information to different decision makers. Further, these systems can operate at a variety of levels, from the international to the local. For example, a national system may deliver information to the national government, a local system would deliver information to a local community, and a corporate system would deliver information to the managers of an enterprise.

For convenience, we will generally refer to 'the' informing system, and often use national scale systems as the main source of examples. However, the principles described apply equally to systems at other scales, and attention will be given to local and corporate systems as well as those that serve national governments.

For analytical purposes, we can identify the elements of an informing system as: the broad conceptual framework, the interpretative context, theories, models, tools, predictive models, institutional mechanisms, targets and benchmarks.

INFORMING SYSTEMS IN FLUX

The central thesis of this book is that informing systems are evolving from a 'traditional' to an 'emerging' form. Table 1.1 summarises the characteristics of the 'traditional' and 'emerging' systems as well as the characteristics of informing systems as they are now.

Table 1.1
Traditional, current and emerging informing systems

	Traditional system (20 years ago)
Broad conceptual framework	Economic growth with environmental remediation
Interpretive context	Economic growth is the main concern of decision makers
Theories	Taken from neo-classical economics. Environmental and social sciences relatively poorly developed and/or uninfluential.
Models	The economy operates with unlimited natural resources.
Tools	Economic indicators, limited range of social indicators.
Predictive models	Economic predictive models highly developed and widely influential.
Institutional mechanisms	Highly developed institutional mechanisms for gathering and analysing economic and some social data. Limited environmental data gathered, mainly to protect human health.
Targets and benchmarks	Economic targets drive policy. Social and environmental benchmarks are limited, and mostly the bare minimum to maintain stability and protect human health.

The 'traditional' and 'emerging' informing systems described here are cartoons. They illustrate salient features of past, current and potential future systems without pretending to describe accurately the real features of any. Accordingly, we do not suggest that the system in place 20 years ago was the 'traditional' system described here, or that the system in place 20 years hence will be exactly the same as the 'emerging' system. However, we do contend that present systems draw many of their salient features from the 'traditional' system, but are progressively incorporating more characteristics of the 'emerging' system.

Current system	Emerging system (20 years from now)
Decision makers aware of sustainable development, but many operate in traditional mode.	Sustainable development
Increasing community awareness of environmental and social issues, and greater political and corporate commitment to them. Economic concerns remain central.	Economic, social and environmental considerations are all-important and are given appropriate weight.
Environmental and social sciences becoming increasingly sophisticated and influential.	Powerful and influential theories for understanding the environment and society as well as the economy.
A number of models are emerging, (for example, PSR, ecosystem health, resource economics).	It is still not clear which models will be useful.
Emerging tools include environmental indicators, 'green accounting' and sustainability indicators.	It is still not clear which tools will be useful.
Economic predictive models continue to be very influential. Some environmental predictive models being developed and beginning to influence policy making.	Environmental predictive models as well developed and widely used as economic predictive models.
Institutional mechanisms for reporting on the environment are embryonic. Few institutions are responsible for acting on the basis of environmental information.	Strong institutional base for gathering and analysing economic, social and environmental data in an integrated fashion. There are institutions responsible for acting on the basis of social, economic, and environmental information.
Environmental targets and benchmarks are emerging, but in an ad hoc manner and often resisted by vested interests.	Environmental and social targets are as important as economic targets.

THE 'TRADITIONAL' SYSTEM

The distinguishing characteristic of the traditional system is its strong emphasis on economic factors. This reflects a view that economic management is the major role of government[2] while making a profit is the major task of corporations. The traditional system's *conceptual framework* assumes that there will be no major environmental or social problems. The capacity of the natural environment to assimilate wastes is regarded as unlimited, and natural resources are considered to be either inexhaustible or readily replaced by a practical equivalent when they are exhausted. Social problems are thought to be kept in check by a stable, relatively homogenous society based on the nuclear family.

It is further assumed that increased economic prosperity will strengthen social cohesion by reducing pressure on families and individuals — thus reinforcing the primacy of economic considerations. Any social or environmental problems that arise can be remedied by ad hoc and relatively inexpensive measures. For example, relief can be offered to communities affected by natural disaster or regional downturn, welfare will assist families or individuals temporarily experiencing difficulty, and accidents (such as oil spills) with adverse environmental effects can be 'cleaned up'.

Whereas the conceptual framework refers to the intellectual milieu within which decisions are made, the *interpretative context* is the attitude or orientation of decision makers and those who surround them. In line with the preceding paragraphs, the *interpretative context* of the traditional system is a strong emphasis on economic growth as the most important outcome for which governments and other institutions ought to strive. In the private sector this corresponds to maximising profits and/or income. This attitude does not necessarily represent a devaluation of society or the environment, but rather reflects an underlying assumption that they will 'take care of themselves'.

Not surprisingly in view of this emphasis on economics, the technical apparatus required to process and interpret economic data are well developed and influential. By technical apparatus we mean the *theories, models and tools*[iii] that are used to guide the selection, processing, and interpretation of data and information. Importantly, these theories, models and tools reflect the assumptions of the broad conceptual framework. For example, economic theories tend to assume unlimited natural resources, and methods for deriving key economic indicators do not take social and environmental factors into account. The apparatus themselves may vary: neo-classical economic theories currently hold sway, but Keynesian views were in vogue as recently as the 1970s.

Theories, models and tools for studying social and environmental trends, on the other hand, are less well developed. A limited set of social measures is in place — mainly related to demography and income distribution. Environmental tools and models tend to be closely linked to human health. That is, the environment is seen as being important to the extent that it affects human health, and tools for monitoring the environment concentrate on potential damage to human health (for example, air pollution, heavy metals in drinking water).

In the traditional system, *predictive models* for the economy are well developed, widely used, and have a strong influence on decision making. For example, Treasury and private forecasters regularly make projections of economic growth, inflation, unemployment, current account deficit amongst others, and these are used to guide government policy and private investment decisions. Limited social modelling takes place, but this is largely restricted to studies of demographic changes that can be used to plan infrastructure investments and delivery of key services.

The traditional system is also marked by very well developed *institutional mechanisms* for economic information. For example, in most developed nations, a central statistical organisation is responsible for generating key economic indicators — including gathering raw data and processing those data into the required form. In the case of the national accounts, the United Nations' System of National Accounts provides international guidelines to which most national agencies adhere. Typically, the central statistical agency will have a statutory mandate to produce such indicators, possibly with enforcement powers to ensure data are provided in a timely manner.

Traditional systems also have institutions responsible for responding to economic information. Generally, treasuries, finance departments and central banks keep a close watch on economic information, and respond to changes by adjusting fiscal and monetary policy in 'standard' ways. For example, it is common practice to respond to sluggish growth in the national accounts by reducing official interest rates.

The institutional machinery for generating environmental data is much less developed in traditional systems. Collection of environmental data is typically ad hoc, with little consolidation or integrated analysis. The links between decisions and information are generally poorly developed. The few environmental data collected are often used to protect human health and not the broader environment. For example, there may be minimum standards for the concentration of certain pollutants in air and water, and a particular agency or government department may be responsible for monitoring compliance with these standards.

As for the other aspects of the traditional system, *targets and benchmarks* are well developed for the economy, but virtually absent for the environment. Central banks are committed to formal or informal inflation targets. Governments are committed to a target for economic growth as measured by the national accounts, which is in the range of 3–5 per cent for developed nations and around 8–10 per cent for developing countries. International comparisons are another form of benchmark for economic indicators.

THE 'EMERGING' SYSTEM

The 'emerging' system is our vision for a system that will meet the information needs of a sustainable society. Sustainable development is the *conceptual framework* for the emerging system.

All versions of the sustainability concept involve a long-term perspective and a conviction that it is necessary to take full account of environmental and social as well as economic factors when making decisions. The emerging system will operate in a society that has translated the concept of sustainable development into a set of social and political priorities that gives due prominence to environmental, social and economic outcomes. This will be expressed in explicitly recognised economic, environmental and social goals and a requirement that any decision must promote all three. This approach will provide the *interpretative context* for the emerging system.

The *theories, models and tools* of the emerging system will be suited, where appropriate, to an integrated analysis of economic, social and environmental trends and outcomes within the sustainability paradigm.

In most cases, integrated analysis of information from the economic, social and environmental spheres will probably be carried out using flexible models rather than a single theory or tool. A theory capable of integrated analysis would involve a precise, purely objective, formulation of the relationships between critical aspects of society, the economy and the environment. Given the subjective and qualitative nature of aspects of sustainable development and the current state of theories relating the environment, the economy and society, this seems unlikely. For similar reasons, a single tool (such as an aggregated index) that relates all aspects of sustainable development is unlikely to be meaningful.

The economic theories and tools used in the emerging system may be different to those employed in the traditional system. One school of thought suggests that traditional economic theories and tools must be modified in order to take environmental and social factors more fully into account (Daly and Cobb 1989, Eckersley 1998). This is discussed more fully in Chapters 4 and 6.

In the emerging system, *predictive models* will be available to build pictures of economic, social, and environmental outcomes under a range of scenarios. It is unlikely that a single model will be able to project outcomes for all aspects of the environment (climate change, air pollution, soil loss and so on), let alone model economic and social outcomes as well. A suite of predictive models that can be used in concert to explore broad scenarios is more feasible. In such an arrangement, the outputs of one model would constrain other models. For example, a model showing the effects of increasing greenhouse gas emissions on world climate might constrain the amount of fossil-fuel energy that could be used as an input to a model of the economy.

The *institutional mechanisms* in the emerging system will also reflect the greater emphasis on social and environmental information. Changes will be greatest for institutions collecting environmental data. Their activities will be better focused using the tools, models and theories noted above and they will have a more secure basis and more resources for their work.

A second change from the traditional system will be that institutions are designed to integrate economic, social and environmental information rather than being concerned solely, or principally, with only one of these. It is only a slight caricature to suggest that in the traditional system the institutions responsible for economic, social and environmental management are seen as adversaries or competitors. In the emerging system, these adversarial relationships will be replaced by co-operative arrangements. Integrated institutions having broad responsibility for sustainable development rather than the economy, welfare, or environmental protection are also possible, although there may be administrative advantages to retaining organisations specialising in particular aspects of sustainable development.

The third characteristic of institutional mechanisms in the emerging system is that there will be institutions responsible for responding to 'signals' from the environment and society as well as the economy.

This leads naturally to the observation that in the emerging system, *targets* and *benchmarks* will be set for environmental and social as well as economic parameters. Economic, social and environmental targets will be set and analysed in concert rather than individually. Targets and benchmarks will be based on explicit goals and objectives, and should be constantly revised in light of changing circumstances. This is particularly important with environmental goals and targets, in view of the imperfect knowledge of how many systems operate, and in keeping with the ethos of 'adaptive management'.

DIRECTIONS OF CHANGE

As stated above, the thesis of this book is that informing systems are changing from something like the traditional system to something that more resembles the emerging system. The chapters give more details on these changes and the possibilities for the future.

The changes described here are not consciously co-ordinated or planned, nor has the pace of change been uniform in different places, spheres of activity, or components of the informing system. Rather, change is driven at varying paces by broader historical and other forces.

CONCEPTUAL FRAMEWORK

The last decade has seen increasing acceptance of the concept of sustainable development, sparked by the publication in 1987 of the Brundtland Report (WCED 1987) by the World Commission on Environment and Development. In response to this report, the United Nations established the Commission for Sustainable Development, and encouraged countries to develop national sustainability strategies. Many nations have already done so. Sectors, local governments, and corporations are also developing sustainable development strategies.

The concept of sustainable development is relatively easy to state in the broad, but notoriously difficult to apply in particular circumstances. A great deal remains to be done to 'translate' the concept of sustainable development for application to different enterprises and organisational units and at a range of spatial scales. Governments, communities, commercial sectors, businesses, and individuals must understand what it means for them to contribute to sustainable development in their own particular, dynamic and contingent circumstances.

The evolution of the sustainable development concept is discussed more fully in Chapter 2, with an analysis of how Australian governments (and some industry sectors) grappled in the 1990s with 'translation'.

INTERPRETATIVE CONTEXT

There has been a trend in the last two decades toward greater emphasis on environmental issues. Surveys of public opinion in Western democracies consistently show that the environment is a major concern, and respondents often rank it with economic development. At the same time, the environment has shifted from the fringe of political debate and become a mainstream issue. Recent statements by senior officials in the Treasury and the Reserve Bank acknowledging the importance of environmental considerations when planning for Australia's economic future are further signs of a changing interpretive context. This does not mean, however, that environmental, social and economic considerations are now equal partners in the political sphere. Economic issues still dominate.

THEORIES

The developments in theory most important for evolving informing systems are taking place in ecology and economics. Ecology is a relatively young science. It emerged as a distinct discipline only at the beginning of the twentieth century, and did not grow significantly, at least in terms of numbers of practitioners, until the middle of that century (Bowler 1992). The functioning of ecosystems is critical in any consideration of sustainable development. The theories that provide the language in which ecosystems can be described and analysed are still being refined, and ecologists have yet to discover how many of the world's ecosystems work. As an example of the rapid development of this field, the concept of biological diversity — critical in most accounts of sustainable development — was clearly stated only in the last 20 years.

It has been long recognised that economics, as practised through most of its disciplinary history, does not adequately take the environment into account. Recent decades have seen various attempts to correct this, through sub-disciplines known variously as 'environmental economics', 'ecological economics', or 'resource economics'. A number of partial solutions have been found, and economics is much better placed to contribute to environmental debates than it was 30 years ago.

Whether economics can ever take full account of all environmental considerations is debatable. Perhaps it should not be expected to. Other disciplines provide valid but limited insights, and why should we suspect that economics is any different? In the meantime, there is considerable debate over the degree to which economics can, or should, be 'adjusted' for environmental concerns. The argument over whether the technique of 'discounting' should be applied to the environment is a case in point. The to and fro between advocates of 'strong' and 'weak' sustainability is another. 'Strong' sustainability effectively quarantines some aspects of the environment from economic analysis. Developments in economic theory are discussed more fully in Chapter 4.

MODELS

In recent years a plethora of models has sprung up to help us think about sustainable development, or aspects of it. The models that have received the most attention include the pressure-state-response model developed by the OECD for reporting on the state of the environment, the concept of ecosystem health and various sustainability frameworks that set out criteria for the sustainability of a particular activity or sector. In the corporate sector, the natural step model has won many followers, as have various management frameworks which, while not models in quite the same sense, can be used in a functionally equivalent way to organise information.

It is unlikely that all of the models will survive the coming decades. However, at this stage it is not clear which models will best help decision makers think about sustainable development. The current situation is probably best viewed as a period of healthy competition between rival models, the 'fittest' of which will survive and be used in the future. Chapter 3 discusses these emerging models in more detail.

TOOLS

As with models, a wide range of tools to provide information relevant to sustainable development is being developed. Environmental indicators are designed to measure environmental trends in the same way that economic indicators measure economic trends. Sustainability indicators are related to environmental indicators, being suites of economic, social and environmental indicators arranged in a suitable framework. These indicator approaches typically rely upon independent measures of different aspects of sustainability, allowing decision makers to weigh the various elements.

Other tools are based on modifications of economic theory that take social and environmental factors into account. Adjusting gross domestic product so that it reflects social and environmental factors is a common strategy. There are several approaches to such a modification, ranging from constructing 'satellite accounts' for the environment that stand alongside the 'standard' GDP, to various so-called 'Green GDPs' that assign monetary values to social and environmental factors.

Another approach is 'quality of life' indices, which combine economic, social and environmental measures. The Human Development Index is the best known of these. Which of these tools will be most useful to decision makers, and which will survive in the long term, is still an open question. The emerging tools are described more fully in Chapters 5, 6 and 7.

Environmental impact assessment (EIA) is a process for systematically assembling information about the environmental impacts of a project or suite of projects and has provided useful input to many decisions touching on sustainable development. Although not a tool in the sense considered here it is touched upon in Chapter 9.

PREDICTIVE MODELS

Economic and demographic models have had a powerful influence on decision making for many years. Possibly the only environmental predictive models to have a comparable influence is the suite of models relating to global climate change, which played a major role in securing global action to reduce greenhouse gas emissions. Global climate change models were accepted only after a long struggle to convince decision makers of their value.

The controversy that surrounded these models was instructive. Decision makers found it difficult to act on the projections produced by the models. While it is not possible to say with certainty why this was so, a number of factors may have played a part. Lobbying by vested interests is an obvious one, but more subtle considerations are probably also relevant.

Most decision makers are unfamiliar with the nature and uncertainty of scientific models. Whereas the uncertainty of economic projections is taken for granted (as witnessed by the biannual revision of Treasury forecasts) decision makers criticise environmental models if their outputs are uncertain. Similarly, the fact that different economic forecasters make widely varying predictions about the economy does not reduce the confidence of decision makers in economic predictive models, but disputes amongst scientists about the magnitude of climate change and its effects led to grave reservations on the part of decision makers about the wisdom of acting on global climate change models. Of course lobbyists exploited uncertainty in scientific models, so the two factors reinforced one another.

Hopefully, the breakthrough achieved with climate change models will presage greater acceptance of environmental predictive models as decision making and policy tools. At the same time, it is necessary to develop environmental predictive models of greater sophistication and certainty. Advances in the understanding of ecosystem functioning and improvements in computer technology are combining to achieve this.

Chapter 8 presents detailed examples of some models that have recently been developed by the CSIRO to help with decision making and policy analysis.

INSTITUTIONAL MECHANISMS

All developed nations and many developing nations have now established institutional mechanisms for reporting on the environment to parallel existing arrangements for reporting on the economy and on social trends. The most common mechanisms are state of the environment reporting, supported by the OECD and the United Nations Environment Programme, and national statistical agencies producing 'satellite' environmental accounts, based on the United Nation's System of National Accounts.

However, these institutional arrangements are fragile compared to those for economic and social reporting. While institutional arrangements for generating and bringing together, for example, information about the national accounts are cohesive, comparable environmental data are scattered, or non-existent, and difficult to integrate.

Government processes have been established in some countries and international organisations to examine the sustainability of particular sectors. The Montreal Process on Sustainable Forest Management recently involved 12 nations in a first approximation report on key information about the sustainability of forest management. A few countries have begun to explore the possibility of developing and reporting on a set of national sustainability indicators, or even a 'sustainability index'. In 1996 the United Kingdom released a draft set of 120 sustainability indicators, followed in 1998 by a discussion paper suggesting a suite of just 13 indicators.

At the local scale, a number of projects by councils or communities around the world are considering reporting regularly on a set of indicators of the environment or the sustainability of the local area. The Sustainable Seattle project has served as an exemplar for many communities, and Local Agenda 21 — a spin-off from the 1992 Rio Conference — is gaining worldwide popularity.

As mentioned above, corporations, too, are taking a greater interest in reporting on their environmental as well as economic performance. Corresponding organisational changes are required in order to integrate, analyse, and present this information. While these signs are promising, it is not yet evident that the information produced by these institutional mechanisms always penetrates to the core of the decision and policy-making process. The often-tenuous nature of links between information and decision makers is a key weakness of many informing systems dealing with sustainable development. Correcting this will require further changes to the institutional machinery, to ensure that institutions are responsible for responding to environmental and social signals as well as to economic ones. Institutional mechanisms are discussed more fully in Chapter 9.

TARGETS AND BENCHMARKS

The trend towards setting environmental targets and benchmarks has been less pronounced but there are some encouraging signs. The Montreal Protocol on substances that deplete the ozone layer and the Kyoto Protocol on greenhouse gas emission are both examples of international agreements that involved parties accepting binding targets to reduce the production of substances that damage the environment.

Individual nations have independently established benchmarks or targets for environmental parameters that are of national or regional concern. Most developed countries have targets or benchmarks for the concentration of a range of water or air pollutants — although these have often been set to protect human health rather than for purely environmental reasons. However, targets are being set in many countries to achieve environmental objectives not directly related to human health. Australia has adopted formal targets for the preservation of various types of forest ecosystems (Commonwealth of Australia 1992). Concern over the collapse or potential collapse of fisheries has led most countries to set limits on commercial fish catches. Corporations have also set targets for aspects of their operations such as emission of pollutants and greenhouse gases. Targets and benchmarks are discussed more fully in Chapter 5.

CONCLUSIONS

Some broad conclusions may be drawn from the material presented in this book.

- Systems to inform sustainable development are still emerging;
- There are a variety of approaches to the different aspects of emergings systems and the validity of these approaches is often contested;
- Many of the solutions offered are partial, rather than complete;
- There is no single system, or type of system, for informing sustainable development.

Despite these caveats, the quality and quantity of information available to make decisions about sustainable development, and our capacity to make good use of it, is increasing. There is good reason to believe that this trend will continue.

Perhaps these conclusions should not be surprising. When stated in the most general terms, sustainable development is a powerful, intuitively appealing concept. Few would argue with, for example, the Brundtland definition. But the more this concept is focused on particular sectors, groups, or activites, the more difficult it becomes to agree on a meaning. Not only is there debate about meanings, but the seemingly simple unitary concept of 'sustainable development' shatters into myriad concepts such as 'sustainable agriculture', 'sustainable fisheries', 'sustainable steelmaking' and 'sustainable communities'. While these concepts may be the best efforts of stakeholders to 'translate' sustainable development into their own sectors or spheres of life, all too often it is difficult to recognise any familial resemblance.

The various 'translations' are not only due to the fact that different groups of people have been responsible for the translations. There are real differences between different sectors and spheres of life, which dictate differences in the practical meaning of sustainable development in each. It has therefore been necessary to develop many different models, tools and approaches to institutional arrangements, as 'one size' is unlikely to fit all.

A further consideration is that, although sustainable development is an integrative, holistic concept, the tools available to analyse it in detail are derived from a disciplinary and fragmented approach to knowledge. It hardly need surprise us, then, that approaches that are useful for analysing a carefully defined part do not provide a full account of the much more complex whole. Until somebody does come up with a 'theory of everything', which is unlikely, we will just have to accept that science and economics will only give us partial insights into the problems associated with sustainable development.

A concomitant to this is that there will always be a role for subjective judgements in making decisions about sustainable development. The challenge is to make these informed and rational judgements. That is the role of the informing systems that are the subject of this book.

NOTES

1   The material in the first half of this section relies heavily on Bowler (1992), McCormick (1992) and Ponting (1991).

2   Functions such as national defence, maintaining law and order, international relations, education etc are also important, but economic management has an overarching importance and is the main driver of information needs. Matters such as education and infrastructure may be seen to derive much of their validity from their contribution to economic prosperity.

3   There are shades of meaning in these terms. A *theory* refers to formally posited relationships between various entities, often expressed in mathematical formalism. *Models* are a class of less precise intellectual objects, which are best thought of as mental props or extended metaphors. They present a way of thinking about a problem or issue without setting out the relationships between entities as precisely as theories do. *Tools* are specific procedures for transforming data into information and communicating it to decision makers. Examples of tools in this sense are the various indicators (such as GDP, CPI) in common use and the algorithms used to derive these indicators from the raw data on which they are based. Discursive reports can also be viewed as a species of tool in this sense.

2
DEVELOPMENT OF SUSTAINABILITY CONCEPTS IN AUSTRALIA

DAVID BENNETT

INTRODUCTION

At least since Plato lamented in his *Critias* the 'consequences of excessive logging and grazing in the mountainous region of Attica, near Athens' (Coates 1998) and observed that the mountains were 'only the bones of the wasted body' (*Critias*), humans in Western intellectual traditions have been aware that human activity has impacts on their environments. A hemisphere away and nearly 2500 years later, humans are still aware of those impacts. For example, while much of Australia's biodiversity remains in good shape, 'About 40 per cent of Australia's forests have been cleared in the 200 years since European settlers arrived, with another 35 per cent affected by logging' (SEAC 1996). Since 1788, 'five per cent of higher plants, seven per cent of reptiles, nine per cent of birds, nine per cent of fresh-water fish, 16 per cent of amphibians and 23 per cent of mammals are extinct, endangered or vulnerable. Twenty species of mammals, 20 bird species and 68 plant species are known to have become extinct' (SEAC 1996).

For the most part of time between Plato and the present, the impact of human activity on the environment has not overly concerned most humans in Western traditions, because of three very simple assumptions, among others. First assumption: humans in the Western traditions have since the earliest times taken it as part of their worldview that they have a 'right to dominate nature, and to multiply [their] species … Whatever changes have come about in the rest of [their] attitude to the world, dominion and multiplication have persisted and have indeed been intensified. The result of this view of nature as subordinate to man's requirement has been to set man apart from nature' (Black 1970). Second assumption: 'humans have considered themselves above, and immune from, the ecological laws which dictate the numbers and fates of other species' (Happold 1995). Third assumption: there is 'a sufficiency of natural resources to provide the human race with an unlimited supply of wealth' (King 1998).

When human impact on the Earth became obvious or enduring, malpractice, rather than the assumptions, was blamed. In his *Critias*, Plato asserts malpractice by implication. He remarks that Attica before it was overlogged and overgrazed was 'cultivated, as we may well believe, by true husbandmen, who made husbandry their business'. Even when malpractice was shown, it was blamed at the local level and not at the systemic level. Slowly over centuries, then more rapidly over the past five or six decades, these assumptions have been challenged. As human understanding of anthropogenic impacts on the Earth improved, the truth of these assumptions has been increasingly questioned.

A number of factors have been identified as contributing to environmental degradation and have lead to a different appreciation and understanding of relationships between humans and their environment. These factors can be summarised in either or both of the following formulae:

Ed (environmental damage) = P (population) x C (consumption per capita) x D (environmental damage per unit of consumption).

And ...

I (impact on the environment) = P (population) x A (per capita affluence) x T (damage done by technologies supplying each unit of consumption). (Tickell 1997, 455).

Yet, even when there is agreement that the environment is changing and agreement that the formulae capture some notion of the trends of change, there is disagreement about how to interpret these trends. Economists like Julian Simon hold, 'Almost every trend that affects human welfare points in a positive direction, as long as we consider a reasonably long period of time and hence grasp the overall trend' (Simon and Myers 1994). Simon holds, for instance, on the human population aspect of these formulae, 'The doomsayers of the population control movement offer a vision of limits, decreasing resources, a zero-sum game, conservation, deterioration, fear, and conflict ... Or should our vision be that of those who look optimistically upon people as a resource rather than as a burden — a vision of receding limits, increasing resources and possibilities, a game in which everyone can win ...' (Simon and Myers 1994). Simon was noted for engaging in debates and bets with well-known environmental advocates, such as Norman Myers and Paul Ehrlich to prove that although some concern was justified there was no resource crisis, since 'technology would ... find alternatives to existing processes and use of resources when they were needed' (Yencken and Wilkinson 2000).

If Simon represents the anthropocentric end of a spectrum on interpreting these trends, then Deep Ecology represents the ecocentric end of this spectrum. Deep Ecologists characterise the anthropocentric end as 'shallow ecology'. Simon maintains that environmental issues should be judged within the context of human and economic worth, to illustrate, he says, 'that just about every important measure of human welfare shows improvement over the decades and centuries' (Simon and Myers 1994). This view is clearly set in an anthropocentric context.

Deep Ecologists reject 'the assumption that humans and human projects are the only items with value' as well as 'the assumption that humans and human projects always outvalue other considerations and the value of other things' (Sylvan and Bennett 1994). Arne Naess, the founder of Deep Ecology, contrasts the positions, 'The shallow ecology movement talks only about resources of mankind, whereas in Deep Ecology we talk about resources for each species' (Bodian 1982). On the human population, Deep Ecologists maintain, 'The flourishing of human life and cultures is compatible with a substantial decrease of the human population. The flourishing of non-human life requires such a decrease' (Naess and Sessions 1984). This, of course, does not mean draconian measures to eliminate current humans, but sensible policies to slow population growth. On technology and wealth, Deep Ecologists hold, 'Present human interference with the non-human world is excessive, and the situation is rapidly worsening' and an ideological change to 'appreciating *life quality* (dwelling in situations of inherent value) rather than adhering to an increasingly higher standard of living' (Naess and Sessions 1984).

There is no intention of attempting to resolve this debate here. The intention here is to examine the concept of sustainability, which embodies responses to the above formulae and the sorts of issues raised by such debates. Despite disagreement over interpretations and over the paths, both ends of the spectrum would agree, 'sustainability is the outcome we desire' (Yencken and Wilkinson 2000).

Although sustainability is a broad concept, a good deal of the recent discussion of the concept has centred on 'sustainable development' and 'ecologically sustainable development (ESD)', which in turn have been multifariously defined. Two Chairs of Australia's Ecologically Sustainable Development Working Groups, Stuart Harris and David Throsby state, 'a review of the literature revealed nearly 300 definitions of sustainable development' (Harris and Throsby 1998). So, to state the obvious:

Sustainable development may mean different things to different people, but the idea itself is simple. We must work out models for a relatively steady state society, with population in broad balance with resources and the environment. (Tickell 1997, 456).

Sustainability recognises two key tenets: 1) needs and 2) limitations in the broad balance between development and the environment. The single, most often quoted definition of sustainable development comes from the World Commission on Environment and Development (WCED) formulated in *Our Common Future*, also known as the Brundtland Report after Gro Harlem Brundtland, the head of the Commission. The WCED definition has become a reference point for meanings of sustainable development. This definition brings out these two tenets:

> Sustainable development is development that meets the needs of the present without compromising the ability of future generations to meet their own needs. (WCED 1987, 87).

While the statement of needs is clear, the statement of limitations is less so. Recognising that future generations have needs and that those needs will have to be met places limitations on the rate and methods of the current generation in meeting their needs. One limitation is temporal. The limitation on the present generation is to develop in such a way that does not exhaust a resource and exclude its use or enjoyment by future generations. Or put another way, a limitation on natural resources, even on natural resources that are in theory renewable, is that they should not be used more rapidly than they can be replaced. Despite the pervasiveness of the WCED definition, there is a strong impression that sustainable development has acquired a use without having acquired any clear meaning. Herman E. Daly expresses colourfully the consequences of not providing sustainable development with a clear meaning, 'call it "sustainable development" — in the hope that chanting this mantra will free us from the obligation to define it, and absolve us from our addiction to robbing the future' (Daly 1992).

One of the outcomes of the 1994 Fenner Conference on the Environment, *Sustainability: Principles to Practice* states 'the best approach to the problem of understanding the concept of sustainability may be to encourage the various sectors and interest groups in Australia to translate the principles into their own languages and contexts as a basis for implementation' (Harding 1996). The conference outcomes held that the 'language of ESD' had become a barrier to implementation of the principles of sustainability. These principles will be taken up later. If the recommendation that the various sectors should express these limitations in their own language were followed, then sectors as defined by the Commonwealth of Australia (1992), such as agriculture, fisheries ecosystem management, forest resource use and management, manufacturing, mining, urban and transport planning, tourism and energy use, energy production and transport would not only express them in different ways, but also would provide for the needs of the future in different ways. For example, a sustainable agriculture would be a whole-systems approach to food, feed and fibre production that balances environmental soundness, social equity, and economic viability among all sectors of the public, including international and inter-generational peoples. Inherent in this is the idea that sustainability must be extended not only globally but also indefinitely in time, and to all living organisms including humans.

Establishing a reference point and recognising key tenets is a useful, yet insufficient step in interpreting sustainability, and comprehending the development of the concept. The meanings of sustainability are as much a function of historical evolution as interpretation.

A HISTORY OF ELEMENTS OF THE CONCEPTS OF SUSTAINABILITY

While the terms 'sustainable development' and 'ecologically sustainable development' are recent, dating back only two or three decades, the concept of sustainability can be traced over centuries. There are several important and often mentioned seminal developments in this history.

STEWARDSHIP

In Western Judeo-Christian traditions, the concept of sustainability is as old as the concept of stewardship. Stewardship involves looking after something, such as land or natural resources, and taking care of it, without owning it. A steward looks after something on behalf of its owner. The relationship is based on trust — the owner trusts his or her steward to prudently care for his or her possession, use it sensibly and in a sustainable manner, and to give it back in an equal or better condition when the time is right. Stewardship dates back at least to Genesis of the Old Testament of the Bible and to the post-Platonic philosophers of the Roman Empire.

> The tradition of stewardship is derived from a hierarchal arrangement God:Humans:Nature. Under this arrangement God put humans on the earth in order that they should look after it, i.e., nature. While humans served as stewards, the ultimate ownership of the earth was never for a moment in doubt. (Sylvan and Bennett 1994, 70).

Australian philosopher, Val Plumwood argues that 'according to at least some versions of the Stewardship position, humans do not have absolute title to the earth but are merely Stewards for God, and have obligation to care for the plants and animals of the earth because God cares for them, even if humans do not. Thus they are not entitled to manipulate the earth exclusively for their own benefit' (Routley 1975). Attfield describes the point of stewardship thus:

... stewards are essentially managers who act on behalf of owners ... and ... the point of the metaphor is the steward's responsibility and answerability, not the devaluation of the world which is their trust, and which is regarded as a reflection of the divine glory, and judged by its creator to be 'very good'. Even if the tradition is secularized and adopts a nontheistic form, people do not forfeit their responsibilities, but remain answerable to the community of moral agents for the fostering and the preservation of all that is intrinsically valuable. (Attfield 1991, 61).

Hence, the concept of sustainability is a latter day progeny of a stewardship position in that both recognise limitations on the demands that can be placed on the environment, although the origins of those demands may be different. Also both recognise the necessity for maintaining resources because there are needs and interested parties beyond the present generation. In 'meeting the needs of future generations' sustainability replaces the God of stewardship with posterity. This is succinctly phrased in the oft quoted popular expression, 'We don't inherit the earth from our ancestors, but borrow it from our children'.

Secularised stewardship can be understood along the lines that The Natural Step, a non-profit environmental education organisation, sets out in their Four System Conditions:

1   In a sustainable society, nature is not subject to increasing concentrations of substances extracted from the earth's crust. This means that fossil fuels, metals, and other minerals cannot be extracted at a faster rate than they are re-deposited back into the Earth's crust.

2   In a sustainable society, nature is not subject to increasing concentrations of substances produced by society. This means that things like plastics, ozone-depleting chemicals, carbon dioxide, waste materials, etc. must not be produced at a faster rate than they can be broken down in nature.

3   In a sustainable society, nature is not subject to increasing degradation by physical means. This means that we cannot harvest or manipulate ecosystems in such a way as to diminish their productive capacity, or threaten biodiversity.

4   In a sustainable society human needs are met worldwide. This means that basic human needs must be met with the most resource-efficient methods possible, including a just resource distribution. (Adapted from http://www.naturalstep.org).

Current humans have an obligation to society and to the planet to manage the Earth well and pass it on to the next generation without excess pollution, the depletion of resources, the destruction of species and of wilderness and the growth of deserts. This obligation extends to the planet, because:

Stewards certainly are in most cases responsible to owners, but if creation consists of bodies each with their own glory (I Cor. 15:40), it cannot be regarded merely as expendable resources or as disposable property. Most adherents to the stewardship view have implicitly accepted that intrinsic value is to be found among nonhumans as well as many humans; this granted, stewards of the earth should be seen not only as managers of resources, but equally as curators of treasures or as trustees of the biosphere. The property metaphor suggests that nature is regarded solely as instrumental; but on the stewardship view is has characteristically also been regarded as of value in itself. (Attfield 1983, 216–17).

RACHEL CARSON'S SILENT SPRING — 1962

Biologist and writer Rachel Carson's legacy to sustainability was a graphic demonstration of an instance of unsustainability. She supplied an imperative for developing the concept of ecologically sustainability. 'As early as 1961 [sic], Rachel Carson's *Silent Spring* had highlighted the need for more concern to be shown regarding the effects that humankind was having on the environment' (ESD Working Groups 1991: v). She questioned humanity's faith in technological progress and helped set the stage for the environmental movement. Indeed, *Silent Spring* is widely viewed as the beginning of the modern environmental movement.

Much of human food production in developed countries skates on a thin veneer of technology. Carson demonstrated that the profligate use of synthetic chemical pesticides, particularly chlorinated hydrocarbons such as DDT, was causing serious pollution and killing many animals. Specifically, she detailed how the pesticide DDT had entered the food chain and caused major problems for birds at the top of the food chain. Because of DDT, the shells of raptors, such as bald eagles and falcons, and brown pelicans were too thin. Adult birds were crushing their eggs. It was the first time anyone had publicly shown how poisons affect everything in nature. She illustrated the hazards of the pesticide/resistance cycle and bio-accumulation. In doing so, she also extended the obligation to future generations beyond future generations of humans and to future generations of plants and animals as well.

Carson called for a change in the way humankind viewed the natural world, arguing that nature was vulnerable to human intervention and that human beings were but one part of nature distinguished primarily by their power to alter it, in some cases irreversibly. She argued that human beings are a vulnerable part of the natural world subject to the same damage as the rest of the ecosystem.

She illuminated one of the fundamental limitations to development: at times, technological progress is so at odds with natural processes that it must be curtailed. While not all technological change is necessarily unfortunate or requires remedial action, Carson outlined threats — the contamination of the food chain, cancer, genetic damage, the deaths of entire species — too frightening to ignore. For the first time, the need to regulate development to make it sustainable in ecological terms in order to protect the environment became widely acknowledged.

UNITED NATIONS CONFERENCE ON HUMAN ENVIRONMENT: STOCKHOLM — 1972

'At the international level the concept of ecologically sustainable development was first developed in a cohesive fashion at the United Nations Stockholm Conference in 1972' (ESD Working Groups 1991). By 1972, the environmental crisis had come so evident that it could no longer be ignored and the United Nations Conference on Human Environment was convened.

While none of the seven introductory paragraphs nor the 26 Principles of the Declaration of the United Nations Conference on the Human Environment specifically uses the term 'sustainable development', paragraph 2 nevertheless sets out the concept with specific application to humans:

> The protection and improvement of the human environment is a major issue which affects the wellbeing of peoples and economic development throughout the world; it is the urgent desire of the peoples of the whole world and the duty of all Governments.

Paragraph 2 places a limitation on development in terms of human wellbeing. In effect, it acknowledges that the quickest way to drive a species to extinction is to destroy its habitat, or as it is phrased in the case, its environment. Following on from this, the opening sentences of paragraph 6 have a resonance with both the earlier warnings of Rachel Carson and the later Brundtland definition:

> A point has been reached in history when we must shape our actions throughout the world with a more prudent care for their environmental consequences. Through ignorance or indifference we can do massive and irreversible harm to the earthly environment on which our life and wellbeing depend. Conversely, through fuller knowledge and wiser action, we can achieve for ourselves and our posterity a better life in an environment more in keeping with human needs and hopes. There are broad vistas for the enhancement of environmental quality and the creation of a good life.

Paragraph 6 elucidates the two key tenets of needs and limitations in the balance between development and the environment. It elucidates needs in terms of prudent care for an environment more in keeping with human needs. It elucidates limitations in the balance between development and the environment in terms of enhancing environmental quality simultaneously with creating a 'good life'. While the Declaration recognises these two key tenets, it nevertheless makes it clear in Principle 11 that the 'balance' is tipped in favour of development:

> The environmental policies of all States should enhance and not adversely affect the present or future development potential of

developing countries, nor should they hamper the attainment of better living conditions for all, and appropriate steps should be taken by States and international organizations with a view to reaching agreement on meeting the possible national and international economic consequences resulting from the application of environmental measures.

Mitigating the effects of tipping the 'balance' in favour of development, Principle 13 calls for prudence:

> In order to achieve a more rational management of resources and thus to improve the environment, States should adopt an integrated and co-ordinated approach to their development planning so as to ensure that development is compatible with the need to protect and improve environment for the benefit of their population.

It is obvious that the Declaration ties prudence to sustainability. Further it ties rational management to limited development, thus modification without devastation. Going back to the idea that each sector must define sustainability, in doing this, each sector must act prudently and each sector must rationally manage development in a manner that protects and improves the environment.

WORLD CONSERVATION STRATEGY — 1980

> The World Conservation Strategy was published in 1980. It emphasized that humanity, which exists as a part of nature, has no future unless nature and natural resources are conserved. It asserted that conservation cannot be achieved without development to alleviate the poverty and misery of hundreds of millions of people. Stressing the interdependence of conservation and development, the WCS first gave currency to the term 'sustainable development'. (IUCN/UNEP/WWF 1991).

Now published as *Caring for the Earth: A Strategy for Sustainable Living*, the World Conservation Strategy uses 'sustainable development' to mean: 'improving the quality of human life while living within the carrying capacity of supporting ecosystems' (World Conservation Strategy 1991). This interpretation also brings out the idea of a 'good life' as a limitation as well as providing 'carrying capacity' as a rule of thumb in determining sustainability. Carrying capacity is usually defined as the maximum population of a given species that can be supported indefinitely in a defined habitat without permanently impairing the productivity of that habitat. It is important to observe that this definition explicitly states 'within the carrying capacity of supporting ecosystems', because humans normally increase their own carrying capacity by eliminating competing species, by importing locally scarce resources, and by technology. Indeed, trade and technology are often cited as reasons for rejecting the concept of human carrying capacity out of hand.

The World Conservation Strategy therefore emphasised three objectives:

- essential ecological processes and life-support systems must be maintained;
- genetic diversity must be preserved;
- any use of species or ecosystems must be sustainable.
(IUCN/UNEP/WWF 1991, 1).

To be sustainable a strategy must preserve essential ecological processes as well as the species and genetic diversity that make up the components of those processes. These three objectives give substance to a definition of sustainable development. Reflecting back on Rachel Carson's legacy to the meaning of sustainability, these three objectives clearly outline what constitutes unsustainability.

> The aim of *Caring for the Earth* is to help improve the condition of the world's people, by defining two requirements. One is to secure a widespread and deeply-held commitment to a new ethic, the ethic for sustainable living, and to translate its principles into practice. The other is to integrate conservation and development: conservation to keep our actions within the Earth's capacity, and development to enable people everywhere to enjoy long, healthy and fulfilling lives. (IUCN/UNEP/WWF 1991, 2).

By identifying a new ethic of sustainability and the requirements of that ethic, needs and limitations can be translated into practice. Perhaps more important than providing a brief definition, the World Conservation Strategy defines sustainable development as a set of strategies and tools which respond to five broad requirements:

- the integration of conservation and development,
- the satisfaction of basic human needs,
- the achievement of equity and social justice,
- the provision for social self-determination and cultural diversity,
- the maintenance of ecological integrity.

Each of these is a goal in itself and a condition for achieving the others, thus underlining the interdependence of the different dimensions of sustainability and the need for an integrated, interdisciplinary approach to the achievement of development that is sustainable.

OUR COMMON FUTURE — 1987

Despite being the single, most often quoted definition of sustainable development, the lack of concreteness in the Brundtland Report definition and the juxtaposition of such apparently contradictory terms as 'sustainable' and 'development' have engendered many competing interpretations of sustainable development, particularly since the report emphasises that economic growth is needed and advocates a five to tenfold increase, worldwide, in manufacturing output. In case anyone fails to grasp the message, *Our Common Future* states:

> The Commission's overall assessment is that the international economy must speed up world growth while respecting environmental constraints. (p 89).

Yet, much of our existing economic activity is already destroying the natural world around us, as global warming, species extinction, ozone depletion, toxic contamination, rising sea levels, acid rain, and other indicators and events demonstrate. Although Simon would argue that these are short-term effects and that in the long-term economic growth would be good for human wellbeing, as Tickell points out, 'Whether expressed as Ed = PxCxD, or as I = PxAxT, the results as we can foresee them are the same — catastrophe — whether in fast or easy stages' (Tickell 1997).

UNITED NATIONS CONFERENCE ON ENVIRONMENT AND DEVELOPMENT (UNCED) — 1992

The Brundtland Report was the forerunner to UNCED, which was called in response to a growing concern over the environmental degradation of developing countries, vividly illustrated in the Brundtland Report. Both the Brundtland Report and UNCED served to focus greater attention on the Earth's rapidly depleting resources and the need to change the manner in which development is approached, focusing upon sustainable use.

From 3 to 14 June 1992, Rio de Janeiro hosted the United Nations Conference on Environment and Development, better known as The Earth Summit or The Rio Summit or UNCED. The conference was the culmination of two years of negotiations by four Preparatory Committees (PrepComs). With 179 nations in attendance, it was the largest environmental conference ever. For the first time the environment was given equal status with war and economics. The major objective of the Conference was worldwide agreement on environment and development. Or as it is most often stated, sustainable development. The Earth Summit represents the attempt to integrate at the global level the economic side with the environmental side, even if the environmental side came out the lesser quantity. It is an attempt to convert sustainable development from rhetoric into practice.

Five major agreements on global environmental issues were signed. Two of these, the Framework Convention on Climate Change and the Convention on Biological Diversity, were formal treaties whose provisions are binding on the parties. The other three UNCED agreements were non-binding statements on the relationship between sustainable environmental practices and the pursuit of social and socio-economic development. The Statement on Forest Principles pledges parties to more sustainable use of forest resources. Agenda 21 is a wide-ranging assessment of social and economic sectors with goals for improving environmental and developmental impact of each. The Rio Declaration summarises consensus principles of sustainable development.

Principle 4 of the Rio Declaration states:

In order to achieve sustainable development, environmental protection shall constitute an integral part of the development process and cannot be considered in isolation from it.

This statement brings together the key tenets of needs and limitations. Sustainable use cannot be achieved without ecological limitations on development. If development outstrips renewal, sustainability is impossible. If development proceeds in a manner that prevents renewal, sustainability is impossible.

MEANINGS

The major point to be derived from reading the literature critical of the idea of sustainable development is that this is a movement that is more ethereal than concrete. (Davis nd, np).

Providing a meaning or definition for sustainability — or for any term for that matter — can take various forms, but there are two forms of definition that are of particular relevance here: substantive and operational. A substantive definition provides the essence or significance of a term, while operational definition provides information that can be used in decision making. Operational definitions should be consistent with substantive definitions when elaborating the same term. The substantive definition of the Brundtland Report does not indicate how to implement 'development that meets the needs of the present without compromising the ability of future generations to meet their own needs'. If the World Conservation Strategy's five broad requirements do not in and of themselves provide an operational definition, they would provide flexible, yet clear guidelines, if they are taken together with the substantive definition of the Brundtland Report.

The World Conservation Strategy makes the following observation about 'sustainable development':

The term has been criticized as ambiguous and open to a wide range of interpretations, many of which are contradictory. The confusion has been caused because 'sustainable development', 'sustainable growth' and 'sustainable use' have been used interchangeably, as if their meanings were the same. They are not. 'Sustainable growth' is a contradiction in terms: nothing physical can grow indefinitely. 'Sustainable use' is applicable only to renewable resources: it means using them at rates within their capacity for renewal. (IUCN/UNEP/WWF 1991, 9).

This argues that 'sustainable growth' is the oxymoron that is often interpreted to be sustainable development. A nice elaboration of this oxymoron draws on George Orwell's novel *1984*. In *1984*, Orwell describes a society 'in which language is an important means of social control'. The state uses 'Newspeak', the state language, and 'double-think' as means of control. Orwell has a linguistic category B, into which sustainable development interpreted as sustainable growth would fit well. Orwell gives the following description, 'The B vocabulary consisted of words which had been deliberately constructed for political purposes: words, that is to say, which not only had in every case a political implication, but were intended to impose a desirable mental attitude upon the person using them'. They were always compound words, 'A sort of verbal shorthand, often packing whole ranges of ideas into a few syllables'. As sustain, on the one hand, implies 'to keep in being, to cause to continue in a certain state' and development, on the other hand, implies 'change, development or growth from within', the concept of 'doublethink' appears particularly appropriate.[1] Development in other words precludes sustainability. The dialectical tension locked up in sustainable development is easily exposed in other connected ways. In so far as a macro-goal of economics, at least as conceived by development proponents, is increased throughput, and such throughput means enhanced environmental impact, development economics is in diametrical opposition to environmental sustainability.

The inclusion of 'ecological' with 'sustainable development' attempts to identify 'the distinguishing features of an ecological approach to development — taking an integrated approach and taking a long term view' (Lothian 1998). As Andrew Lothian notes a good deal of the debate over sustainable development has centred on achieving balance between the environment and development:

> Commonly the view is expressed that we need to keep these things in balance, that it is a matter of balancing the economic and the environmental ... However in practice it merely provides a publicly acceptable face to a decision which almost invariably favours development over the environment. One never hears the balance argument used when the decision favours the environment — then it is portrayed in terms of a bold new program or a courageous decision. The application of the balance paradigm generally results in win–lose outcomes, wins for development, losses for the environment. Thus the paradigm of balance does not benefit the environment, it merely masks its gradual demise. (Lothian 1998, 54)[2].

The key tenets that need to be incorporated into the meaning of sustainable development are known. A range of the limitations, including classes of activities that must be excluded, has been identified. Having different but consistent operational definitions aligned to a substantive definition can be a benefit rather than a deficit to translating sustainable development into a non-contradictory set of practices. With these in hand it is appropriate to turn to specific definitions relating to Australia.

AUSTRALIAN CONTEXT

For Australian ESD policy there are two fundamental documents: Australia's National Strategy for Ecologically Sustainable Development (NSESD) and the Intergovernmental Agreement on the Environment (IGAE).

The IGAE 'represents the beginning of a new approach to intergovernmental dealings on the environment. It sets out the roles of the parties and establishes the "ground rules" under which the Commonwealth Government, State, Territory and local governments will interact on the environment ... includes broad principles to guide the development of environment policies ... and ...sets out cooperative arrangements on a wide range of specific issues' (Commonwealth of Australia 1992).

The National Strategy for Ecologically Sustainable Development evolved over a number years and is derived from international developments such as the World Conservation Strategy and the Brundtland report. Australia's 1996 State of the Environment report describes its genesis thus:

> Our traditional pattern of economic development has been at question since the publication of the reports of the Club of Rome, the Blueprint for Survival and the World Conversation Strategy, which coined the term 'sustainable development'.
>
> Australian governments adopted the principle of Ecologically Sustainable Development, or ESD, as a major national strategy in 1992, following a national consultative process. (SEAC 1996, 10–4).

The National Strategy was also developed in response to UNCED. Agenda 21 recognised the key role 'played by strategies, plans and policies at a national level', because 'Agenda 21 is not a recipe book of solutions to be followed blindly by each county, irrespective of its particular problems or circumstances' (Commonwealth of Australia 1992). Thus, the NSESD is intended as a co-ordinated and distinctively Australian approach to ESD and recognises 'the significance of potential threats to our environment and economy if we do not take action' (Commonwealth of Australia 1992).

The National Strategy provides a definition of ESD, but the IGAE does not. The NSESD states:

> While there is no universally accepted definition of ESD, in 1990 the Commonwealth Government suggested the following definition for ESD in Australia:

'using, conserving and enhancing the community's resources so that ecological processes, on which life depends, are maintained, and the total
quality of life, now and in the future, can be increased.' (Commonwealth
of Australia 1992, 6).

This definition contains the tenets identified so far. It recognises needs
and limitations, it contains a concept of a good life or in this case 'a quality of life', and the preservation of the ecological processes underlying
sustainability. It also seems to have an element of 'sustainable growth'
when it states that the quality of life can be increased. Only if quality of
life does not depend on increased material consumption does this definition avoid sustainable growth. Only if quality of life 'is taken to be
understood as development (qualitative improvement) without growth
(quantitative expansion) beyond the capacity of the ecosystem (or landscape or region) to regenerate the raw materials extracted into the economy as inputs and to absorb the materials and energy discarded by the
economy as waste' (Lebel and Steffen 1998) is the oxymoron avoided.

Nevertheless one difficulty that remains with a substantive definition that perhaps can only be remedied with an operational definition
is that a substantive definition 'remains too vague to be truly useful as
a guide for human activity because of disagreement on the meaning of
"needs"' or for that matter any other element of the definition (Five
E's Unlimited 1999). A way of resolving this is to set out principles
that while they do not in themselves define needs or the other elements, they do set the parameters outside of which meeting needs
becomes unsustainable. Or put another way, they divide the sustainable
from the unsustainable.

PRINCIPLES

'A principle provides a guiding sense of the requirements and obligations of right conduct' (Harding et al. 1996). An important attempt to
resolve ESD into practical action and right conduct is translating its
elements into principles. Principles that can, if properly implemented,
afford a standardised basis for translating them into the languages and
contexts of the various sectors. The NSESD and the IGAE each contain a set of principles to guide the application of ESD. These principles are in accord with each other as well as with the Rio Declaration,
although the wording of the principles may vary from one document
to another.

The National Strategy sets out the following core objectives and
guiding principles:

The Goal is:
Development that improves the total quality of life, both now and in the
future, in a way that maintains the ecological processes on which life
depends.

The Core Objectives are:
- to enhance individual and community wellbeing and welfare by following a path of economic development that safeguards the welfare of future generations

- to provide for equity within and between generations

- to protect biological diversity and maintain essential ecological processes and life-support systems.

The Guiding Principles are:
- decision-making processes should effectively integrate both long and short-term economic, environmental, social and equity considerations

- where there are threats of serious or irreversible environmental damage, lack of full scientific certainty should not be used as a reason for postponing measures to prevent environmental degradation

- the global dimension of environmental impacts of actions and policies should be recognised and considered

- the need to develop a strong, growing and diversified economy which can enhance the capacity for environmental protection should be recognised

- the need to maintain and enhance international competitiveness in an environmentally sound manner should be recognised

- cost effective and flexible policy instruments should be adopted, such as improved valuation, pricing and incentive mechanisms

- decisions and actions should provide for broad community involvement on issues that affect them.

These guiding principles and core objectives need to be considered as a package. No objective or principle should predominate over the others. A balanced approach is required that takes into account all these objectives and principles to pursue the goal of ESD. (Commonwealth of Australia 1992, 8–9).

An important aspect to emphasise about this list is the final proviso that the 'guiding principles and core objectives need to be considered as a package'. Taken in isolation, a principle may not produce an ecologically sustainable result or worse may work against sustainability. For example, 'the need to develop a strong, growing and diversified economy which can enhance the capacity for environmental protection' produces ecologically sustainable results only if the economy takes environmental impacts into account and only if the economy repays the environment by enhancing the capacity for environmental protection. That is, unless this principle is taken in conjunction with the core objective of protecting biological diversity, maintaining essential ecological processes, and life-support systems, then the result could be 'sustainable growth', which is an oxymoron and unsustainable. This is not to say that the ESD principles are as well understood or as clear as they might be, or that they are fully operational, or even that they are internally consistent.

> Sustainable development was defined by Brundtland as development which aims to meet the needs of people today while conserving ecosystem for the benefit of future generations. While this is reasonably clear, it is actually quite difficult to define and to articulate what it means. The National Strategy sought to explain this by identifying the distinguishing feature of an ecological approach to development — taking an integrated approach and taking a long term view. (Lothian 1998, 54).

The IGAE sets out four principles, elaborated below, that should inform policy making and program implementation:

- inter-generational equity
- the Precautionary Principle
- conservation of biological diversity and ecological integrity
- improved valuation, pricing and incentive mechanisms

The three core objectives and seven principles of the NSESD and the four principles of the IGAE can be discussed under five headings:

- inter-generational equity
- intra-generational equity
- the Precautionary Principle
- conservation of biological diversity
- internalisation of environmental costs

These five match closely the six agreed principles of ESD developed by the Australian ESD Working Groups, which were an important historical part of the development of *The National Strategy for Ecologically Sustainable Development*. The six agreed principles were:

- improving material and non-material wellbeing
- improving equity between generations
- improving equity within the present generation
- maintaining ecological integrity and biodiversity
- dealing cautiously with risk, uncertainty and irreversibility
- taking account of global ramifications of our actions, including international co-operation, international trade and international spillovers. (Harris and Throsby 1998, 7).

The first and last of these taken together can be understood as part of internalising environmental costs.

INTER-GENERATIONAL EQUITY

Inter-generational equity concerns offering future generations environmental quality at least equivalent to that of the present generation. Or put another way, it is ESD between generations. In strictly human terms, the first and second core objectives of the National Strategy provides for 'welfare of future generations' and 'equity within and between generations' and Section 3.5.2 of the IGAE defines inter-generational equity as 'the present generation should ensure that the health, diversity and productivity of the environment is maintained or enhanced for the benefit of future generations'.

This is clearly a stewardship principle. Both of these pick up the Brundtland report's 'the ability of future generations to meet their own needs' as well as Principle 3 of the Rio Declaration, 'The right to development must be fulfilled so as to equitably meet developmental and environmental needs of present and future generations'. Brown Weiss describes this principle to mean:

> Each generation is both a trustee or custodian of the planet for future generations and a beneficiary of previous generations' stewardship. This circumstance imposes certain obligations upon us to care for our legacy just as it gives us certain rights to use the legacy. (quoted in Harding et al. 1996, 9).

The third core objective of the National Strategy can be read to encompass inter-generational equity for non-humans. Taking into account the National Strategy's proviso, protecting biological diversity and maintaining essential ecological processes and life-support systems apply to all species and not merely humans. There is no good reason for limiting ESD to humans and very good reasons within the principles of ESD, such as the conservation of biological diversity, to encompass all species.

> One of the things that we can learn from Aboriginal attitudes to the environment is to appreciate that a good human environment is one that contains all the living organisms it is capable of having. Aborigines would not have been able to conceive of a world without other creatures and plants. A rich life, a full life, is one shared with the rest of the natural word, not one in which everything is subsumed purely to human interests, and where other organisms that get in the way are removed, and where the environment can be damaged uncaringly and causally (Horton 2000, 143).

In human terms, a generation may be 20 or 30 years, but in terms of inter-generational equity, the overlap of generations needs to be considered. Thus inter-generational may be understood as pertaining to the generation who are in a position now to make environmental and developmental decisions considering those who are not in a position to make those decisions as well as those who will come after them.

While determining what constitutes a generation may be difficult to articulate with precision since in chronological terms generations overlap, this conundrum can be sidestepped by defining it in decision-making terms. Determining what constitutes equity is more difficult. 'It is often stated that equity issues are at the core of the idea of sustainable development ... It is claimed ... that the basic objective is equality across generations — we are enjoined not to degrade the environment such that our successors are less well off than ourselves' (Common 1996).

> Equity is not about treating people identically or uniformly ... it is about treating them in a way that provides them with or at least does not deny them equal opportunities, among other things ... Equity is a moral idea, not an assertion of fact. There is no logically compelling reason for assuming that a factual difference in ability between two people justifies any difference in the amount of consideration we give to their needs and interests. *The principle of the equality of human beings is not a description of an alleged actual equality among humans: it is a prescription of how we should treat humans.* (Bennett 1996, section in italics quoted from Singer 1977, 24).

To provide a hypothetical example that will reduce this principle to the simplest terms, apply this principle to a sector, say, fisheries. Fisheries should not make decisions about food production today that will deprive future generations. Overfishing orange roughies could be taken as a case in point both for future generations of humans and orange roughies.

INTRA-GENERATIONAL EQUITY

One reason that many may find the concept of ESD unusable is the simplicity with which some elements of it are stated. For example, the NSESD with regard to intra-generational and inter-generational equity merely states that a core objective is 'to provide for equity within and between generations' (Commonwealth of Australia 1992). The IGAE does not have a separate principle for intra-generational equity.

'Intra-generational equity concerns equity within a single generation. In practical terms it may be taken to refer to equity between the earth's inhabitants at any one time' (Harding et al. 1996). Notions of equity referred to under inter-generational equity apply to intra-generational equity, what is different are that within a given generation not everyone is in an equal position to make decisions about the environment and development.

> A dilemma in considering intra-generational equity is that we have yet to achieve a just world and the evidence of inequality is presently starkly obvious both between and within nations. The possibility that we will achieve a truly equal world within the present generation is

inconceivable, yet this does not render attempts to achieve 'intra-generational equity' doomed, for ... one interpretation of equity is 'unequal treatment of unequals to produce less inequality'. It is this interpretation which should inform the minimum objective in implementing the principle of intra-generational equity. (Harding et al. 1996, 19).

Referring back to the fourth system condition of The Natural Step, policies aimed at equity in distribution would carry intra-generational equity forward. In ethical and economic terms, the central contention in intra-generational equity is the discrepancy between the rich and the poor. Improvement of environmental quality is equally to the advantage of the rich and poor, but the rich are in a better position to pay for it and the poor are in a better position to suffer from poor environmental quality. Take the example of pollution:

> If pollution is a pure public bad then a programme to reduce it will reduce it equally for rich and poor. This does not, of course, mean according to economic criteria that rich and poor are to [be] seen as suffering equally from the existing level of pollution, or that the benefits of reduction will be the same for rich and poor. If environmental quality is a normal, or a luxury, good, the demand for it increases with income. The rich will be willing, and able, to pay more for the given improvement than the poor, and in the standard cost benefit analysis calculus, will be recorded as benefiting more from it. (Common 1996, 10).

The most obvious and direct link between poverty and ESD is, 'Poverty may lead to environmental degradation ... [P]eople who are struggling to satisfy basic needs may not have the luxury of environmental concern' (Harding, Young, Fisher 1996, 20). The economic framework may not be the right framework in which to think about intra-generational equity. 'By characterising the environment as a collection of economic goods that can be put into the consumption basket, economics plays its favourite trick, that of assuming that everything can be compared to everything else using the yardstick of consumer preferences expressed through markets' (Hamilton 1996). Poverty and environmental degradation in terms of intra-generational equity are social justice issues as well as economic issues and moral issues, or in the words of the NSESD guiding principles 'economic, environmental, social and equity considerations'. David Yencken and Debra Wilkinson argue that ecological sustainability is inseparable from social sustainability, economic sustainability, and cultural sustainability. These are the four pillars of sustainability (Yencken and Wilkinson 2000). Harding et al. sum up intra-generational equity in this way: the 'burdens of environmental problems are disproportionately borne by the poorer or weaker members of society' and 'measures to protect the environment may impact on particular sectors of society whilst benefiting another sector' (Harding et al. 1996).

THE PRECAUTIONARY PRINCIPLE

Section 3.5.1 of the IGAE defines the precautionary principle as follows:

1 Where there are threats of serious or irreversible environmental damage, lack of full scientific certainty should not be used as a reason for postponing measures to prevent environmental degradation.

2 In the application of the precautionary principle, public and private decisions should be guided by:

- careful evaluation to avoid, wherever practicable, serious or irreversible damage to the environment;

- an assessment of the risk-weighted consequences of various options.

The second guiding principle of the National Strategy echoes the first paragraph of the IGAE definition.

The precautionary principle is a political, rather than a scientific norm for decisions. The precautionary principle's main thrust is prima facie very simple and straightforward. Where an activity raises threats of harm to the environment, precautionary measures should be taken even if certain cause and effect relationships are not established scientifically. 'The Precautionary Principle is concerned with decision-making under uncertainty ... It is not satisfactory to wait until damage has occurred because it may then be too late to ameliorate the problem' (Harding et al. 1996).

In many ways this is the most controversial of the ESD principles. It 'has been hailed ... as "revolutionary", "groundbreaking", and a fundamentally new approach to environmental management. Others argue it is a dangerous principle that will stop innovation and development' (Harding et al. 1996). The precautionary principle is caught between competing philosophies of life, one eco-centric and risk-adverse, the other more utilitarian and risk-taking. While each of the principles has generated debate, the Precautionary Principle has generated whole conferences such as that conducted by the University of New South Wales in 1993.

The Precautionary Principle offers decision makers a forceful, common-sense approach to environmental problems. It includes taking action in the face of uncertainty; shifting burdens of proof to those who create risks; analysis of alternatives to potentially harmful activities; and participatory decision-making methods. The Precautionary Principle has three major aspects:

- threats of serious or irreversible environmental damage

- lack of full scientific uncertainty

- measures to prevent environmental degradation

For each of these aspects there are arguments that can be mounted about the wisdom or otherwise of using the precautionary principle and what each of these aspects means in terms of practical action. It is obvious, for instance, that lacking scientific certainty about a particular threat could lead to making a decision that exacerbated a problem rather than relieved it. Yet awaiting that certainty may not be an ecologically sustainable option. The litmus test for knowing when to apply the Precautionary Principle is the combination of threat of harm and scientific uncertainty. While the principle is stated in terms of serious or irreversible threats, it could be interpreted to allow for taking action in the face of the cumulative effects of relatively small insults. In the face of long-term negative impact, the Precautionary Principle can be used to obviate imposing environmental degradation on future generations.

The scientific evidence available for a given problem should be taken into account in applying the precautionary principle. Scientific knowledge is a decisive element in the decision whether to adopt the principle or not. Nevertheless, it is only one element and must be weighed against economic, social and ethical aspects.

CONSERVATION OF BIOLOGICAL DIVERSITY

The NSESD and the IGAE are exceedingly terse or succinct on the conservation of biological diversity. The third core principle of the National Strategy for Ecologically Sustainable Development is 'to protect biological diversity and maintain essential ecological processes and life-support systems' and section 3.5.3 of the IGAE states, 'conservation of biological diversity and ecological integrity should be a fundamental consideration'.

The NSESD and the IGAE are backed up by another national strategy specifically dedicated to biological diversity, the National Strategy for the Conservation of Australia's Biological Diversity (NSCABD). The NSCABD begins in the following way:

> Conservation of biological diversity is a foundation of ecologically sustainable development and is one of the three core objectives of the National Strategy for Ecologically Sustainable Development. Biological resources provide all our food and many medicines and industrial products. Biological diversity underpins human wellbeing through the provision of ecological services such as those that are essential for the maintenance of soil fertility and clean, fresh water and air. It also provides recreational opportunities and is a source of inspiration and cultural identity. (NSCABD 1996, 5).

In addition to acknowledging the core objectives and accepting the principles of the NSESD, the NSCABD contains four components in its goal statement. The Strategy recognises that:

- The conservation of biological diversity provides significant cultural, economic, educational, environmental, scientific and social benefits for all Australians.

- There is a need for more knowledge and better understanding of Australia's biological diversity.

- There is a pressing need to strengthen current activities and improve policies, practices and attitudes to achieve conservation and sustainable use of biological diversity.

- We share the Earth with many other life forms that have intrinsic value and warrant our respect, whether or not they are of benefit to us. (NSCABD 1996).

While the NSESD states the need 'to protect biological diversity' and the IGAE recognises the 'conservation of biological diversity and eco-logical integrity should be a fundamental consideration', the NSCABD elucidates this in terms of the integration of 'cultural, economic, edu-cational, environmental, scientific and social benefits' and intrinsic value of ecosystems and other species beyond any use they have to humans. Indeed this last point relates back to the issues of inter-gener-ational and intra-generational equity as it applies to species other than humans. If there were any doubt as to the broadness of this principle, Objective 2.1 of the NSCABD states:

> The development of integrated policies for major uses of biological resources is necessary to coordinate activities within and between all lev-els of government, to ensure that the full social and environmental con-sequences, and the opportunity costs, of development activities are considered, and to ensure that the public interest is properly taken into account.
>
> Integrated policies will also provide the opportunity for all Australians to accept responsibility for the impacts on biological diversity of their activ-ities, including resource consumption, and to participate in achieving ecological sustainability within industries and lifestyles.
>
> Improved management of Australia's biological resources is essential for ecologically sustainable use ... (p 17)

In policy terms, this principle recognises the need for co-ordinated decision making, the potential for applying the Precautionary Principle, and the possible accumulative negative effects of small actions. This principle also recognises what the next and final principle recognises: that it is the human environment, not the human economy that delivers the necessities of life.

INTERNALISATION OF ENVIRONMENTAL COSTS

Core objective 1 and guiding principles 1, 3, 4, 5, and 6 of the NSESD can be translated into the language of internalisation of environmental costs. The IGAE does not define internalisation of environmental costs in those terms. Rather it states the essence of taking environmental costs into account as follows:

3.5.4 improved valuation, pricing and incentive mechanisms —

- Environmental factors should be included in the valuation of assets and services.

- Polluter pays i.e. those who generate pollution and waste should bear the cost of containment, avoidance, or abatement.

- The users of goods and services should pay prices based on the full life cycle costs of providing goods and services, including the use of natural resources and assets and the ultimate disposal of any wastes.

- Environmental goals, having been established, should be pursued in the most cost effective way, by establishing incentive structures, including market mechanisms, which enable those best placed to maximise benefits and/or minimise costs to develop their own solutions and responses to environmental problems.

The concept of 'sustainability' is at root a simple one. It rests on the acknowledgement, long familiar in economic life that maintaining income over time requires that the capital stock is not run down. The natural environment performs the function of capital stock for the human economy, providing essential resources and services. Economic activity is presently running down this stock. While in the short term this can generate economic wealth, in the longer term (like selling off the family silver) it reduces the capacity of the environment to provide these resources and services at all. Sustainability is thus the goal of 'living within our environmental means'. Put another way, it implies that we should not pass the costs of present activities onto future generations. (Jacobs 1996, 17).

Tying internalisation of environmental costs to conservation of biodiversity requires more attention being paid to valuing biodiversity and to applying equitable systems of values. Economic markets fail to recognise the true value of biodiversity, and typically undervalue it. This does not mean putting a price on the head of every item of biological diversity. Instead, biological resources are inadequately regulated and often overexploited. Policy failures, more than lack of scientific knowledge, are usually responsible for the accelerating losses of biodiversity.

One of the more innovative approaches to internalisation of environmental costs is the triple bottom line methodology. John Elkington coined the term 'triple bottom line'.

The sustainability agenda, long understood as an attempt to harmonize the traditional financial bottom line with emerging thinking about the environmental bottom line, is turning out to be much more complicated than some early business enthusiasts imagined. Increasingly, we think in terms of a 'triple bottom line', focusing on economic prosperity, environmental quality and — the element which business has tended to overlook — social justice. To refuse the challenge implied by the triple bottom line is to risk extinction (Elkington 1998, 1).

Sustainability involves the simultaneous pursuit of the interlinked and constantly fluctuating goals of economic prosperity, environmental quality and social equity. Society depends on the economy and the economy depends on the environment. These are interlinked very much as are the five broad requirements of the World Conservation Strategy.

CONCLUSION

Sustainability is an easy concept to grasp but a difficult one to apply. It is founded on tenets and practices as ancient, yet as contemporarily relevant as stewardship. It recognises that humans and other species have needs that must be met at some cost to the environment and ecological processes. There are limitations on the capacity of the environment and ecological processes to fulfil those needs. Sustainability is more than just ecological sustainability. Ecological sustainability is inseparable from social, economic and cultural sustainability. Achieving sustainability encompasses achieving economic prosperity, environmental quality and social equity. Sustainability recognises that a balance between development and the environment must be struck, so that the demands placed on the ecological processes do not seriously or irreversibly damage those processes. In striking that balance, sustainability includes a concept of a good life or a quality of life. The present generation cannot provide itself with its current or an enhanced quality of life at the expense of the ability of future generations to achieve the same or better quality of life as the current generation. Further, that meeting the needs of the present and future generations depends on the preservation of the ecological processes underlying sustainability and that some methods to meet those needs are unsustainable, and are, therefore, unacceptable.

These elements of the concept can be simply summarised in brief statements, such as:

> 'Sustainable development is development that meets the needs of the present without compromising the ability of future generations to meet their own needs.'

> or

> 'A relatively steady state society, with population in broad balance with resources and the environment.'

> or

> 'Using, conserving and enhancing the community's resources so that ecological processes, on which life depends, are maintained, and the total quality of life, now and in the future, can be increased.'

Although succinct and to the point, these statements of sustainability are not the basis for translating sustainability into policy. By converting the concepts into operational principles, sustainability can be incorporated into policies and into decision making. In the briefest form, these principles are: inter-generational equity, intra-generational equity, the Precautionary Principle, conservation of biological diversity, and internalisation of environmental costs. It is through subsidiary policies that sustainability is made operational. It is through sectors translating sustainability into their own terms and applying the principles that the meaning arises. The 1996 Australian State of the Environment report gives this example:

> Sustainable agriculture is based on the principles that: the supply of necessary inputs is sustainable; the quality of basic natural resources is not degraded; the environment is not irreversibly harmed; and the welfare and options of future generations are not jeopardised by the production and consumption activities of the present generation. There is a further objective, which is to maintain or improve yield. Clearly, sustainability is a complex issue that cannot be easily evaluated for modern agricultural systems (SEAC 1996, 6–36).

The next chapters of this book will deal with the meanings of sustainability and the complexities of evaluating the applications of sustainability.

NOTES

1   Borrowed from John Young, 'Sustainable Development: Doublethink of the 1990's.' Talk given on 'Ockham's Razor' *Radio National*, 23 March 1991.
2   Andrew Lothian states that there was an error in his paper about the issue of balance. He states, 'The sentence: "One never hears the balance argument used when the decision favours the environment …". The word "economy" should be substituted for "environment" as it otherwise does not make sense. I have checked the original publication and found it was incorrect also' (pers. comm.) I disagree. It does make sense. Balance arguments rarely favour the environment, so when the environment has a win, it is bold and courageous. For an instance of this see Principle 11 of the Stockholm Declaration.

3
ENVIRONMENTAL MODELS
JOHN HIGGINS

The models discussed in this chapter are conceptual and heuristic schemes for organising, presenting, and analysing information about sustainable development. The scheme is often loose and flexible, to allow for variations in circumstances and purpose. The word 'framework' can be used in the same sense as 'model' but in this chapter the terms are regarded as interchangeable.

In the sciences, 'model' is sometimes used in a more rigorous and precise sense to mean a conceptual representation of a physical or ecological system. By contrast, the models discussed in this chapter do not purport to represent any real physical or ecological system. Rather, they are schemes to help think about, and organise information.

WHAT IS BEING MEASURED?

The term sustainable development, or sustainability or ecologically sustainable development, can be problematic. Like motherhood, it is conceptually simple and universally applauded, but practices vary with circumstances and inclination, and can be divisive. These issues are discussed at length in Chapter 2 and for the purpose of this chapter two important points need to be made. Firstly, models here are based on various 'translations' of the sustainability concept. Variations are partly due to circumstantial exigencies and partially due to broader views and interests of the translators. In this discussion, all translations of sustainability as treated as equally valid. The second, and related, point is that the models described are designed to help organise and interpret information *relevant to* sustainability, although not all deal with sustainability *per se*. Any model dealing with social, economic, or environmental information can be said to be relevant to sustainability but the discussion in this chapter is limited to models dealing with sustainability *per se* and models dealing with environmental information. Models have important benefits for reporting on and monitoring sustainability.

USES AND BENEFITS OF MODELS

SCOPE

Models help define the scope of a reporting or monitoring system. For any particular region, country, sector, activity, or organisation, a range of factors will be relevant to the question of whether development is sustainable. These may include physical, social and economic conditions and processes, various activities, and critical and emerging issues.

A systematic approach to identifying the scope of the reporting or monitoring system is essential. Suitable models provide such a systematic approach, and are also useful for identifying the links and relationships between the various components that must be monitored and reported on.

CONTENT

Once the scope of a reporting and monitoring system has been established, it is necessary to determine what information is required on each issue, condition, process etc. The aim should always be to select the smallest data set that will yield the information needed to make valid judgements and decisions. A suitable model will help by providing a framework for analysing, organising and presenting information. However, the model on its own will not do the whole job. It must be supplemented by specific scientific, social, and economic theories relevant to particular issues and processes.

PRESENTATION AND RELEVANCE

To be effective, a reporting and monitoring system must present information in a way that decision makers can quickly understand, and in such a manner that its relevance is readily apparent. Decision makers will often lack the technical training required to understand scientific, social and/or economic information presented in the style of professionals in these fields. The model can help by clarifying the sort of information that is relevant, and making the context in which it is presented clear. A model, which is understood by both sides, can mediate between decision makers and technical experts.

For example, the pressure-state-response model is widely used for reporting on the state of the environment. While this model has been the subject of considerable criticism, it provides clear guidance on which information is important to decision makers by, for example, identifying human activities that are 'pressures' on the environment.

The best models for monitoring and reporting on sustainability should be simple enough to be readily understood by stakeholders and decision makers; often people without technical training. For this reason, rigorous scientific models, while essential for proper handling and interpretation of information, are generally not suitable for monitoring and reporting. To continue with the previous example, scientists are often frustrated by the pressure-state-response model on first acquaintance, because it seems to imply that the environment is driven entirely by anthropogenic influences. However, this criticism misses the point. Models used for monitoring and reporting are not necessarily intended to be technically rigorous, but to provide a systematic and useful framework within which to organise information and present it to decision makers.

INTERPRETATION

Despite the caveats of the previous paragraph, models for monitoring and reporting can be aids to interpreting data. The important distinction to bear in mind is that this interpretation takes place at a broad, rather than technical level. Models can help answer questions such as 'what does this information tell us about whether the sector/region/enterprise is sustainable?' Reporting and monitoring models do not help answer questions such as 'what is the relationship between fire regimes and biological diversity?' or 'how does consumer confidence influence inflation?'. There are well-documented scientific, sociological and economic theories to answer these questions.

Models vary in how strongly they influence interpretation. Some are little more than a grid within which information is organised. These models provide guidance on what information is relevant to the question of whether development is sustainable, but do not suggest particular interpretations. For example, the 'population-environment-process' model described later in the chapter simply indicates some of the relationships between society, the economy, and the physical environment and suggests what information about these relationships may be useful to decision makers. It is not suggestive of any particular interpretation of sustainability.

Other models are strongly suggestive, and embody a particular view of what sustainable development means. Two examples described below illustrate this. The 'environmental space' model indicates that information about the consumption of natural resources is important, but takes a further step by suggesting that development is sustainable only if resource consumption is kept within certain limits. The 'extended metabolism model' indicates that information must be obtained about 'resource inputs', 'dynamic processes', 'waste outputs' and 'liveability'. It also presents a rubric for determining whether development is becoming more or less sustainable. Reduced inputs and waste outputs with higher 'liveability' correspond to greater sustainability. There are, of course, intermediates between these extremes.

Table 3.1
List of models discussed in this chapter

Focus	Approach	Examples
ENVIRONMENT		
	limiting resources	• ecological footprint • environmental space • sustainable process index • material intensity per service • life cycle assessment
	broad environmental	• pressure-state-response • driving force-pressure-state-impact-response
Increasing focus	environment + services	• ecosystem health • extended metabolism
	environment, social, economic interactions	• driving force-state-response • population-environment-process
	general sustainability criteria	• Natural Step
	specific sustainability criteria	• Sustainability Counts • Montreal process • Local Agenda 21
SUSTAINABILITY		
Management systems	performance measurement continuous improvement	

SURVEY OF MODELS

Many models have been suggested, and used, for monitoring and reporting on either the environment or sustainability. This chapter surveys these models, presents their strengths and weaknesses and suggests circumstances in which each might be useful. Models that have a strictly environmental basis are described first and then the analysis is broadened to include those that deal with sustainability *per se* (see Table 3.1).

Models with an environmental focus are considered in three groups. Firstly, those based on the concept of limiting resources. A subset of these is models based on analysing material flows. Secondly, models designed to present information about a broader range of environmental factors and issues. Finally, there are models that further broaden the discussion of the environment by encompassing the economic and social services it provides.

Models that deal with sustainability *per se* have a similar range and include: frameworks for identifying relevant social, economic, and environmental factors and the links between them; those that are based on general system conditions for sustainability; and those that are based on sustainability criteria designed specifically for the system of interest. Management systems that identify important information, organise that information, and help interpret it are also considered.

MODELS BASED ON LIMITING RESOURCES

The models in this group are designed to draw attention to natural resource limits. Natural resources are interpreted broadly to include biological diversity, and the assimilative capacity of the atmosphere, as well as more traditional resources such as soil, minerals, coal and oil.

These models are highly suggestive of particular interpretations of data, emphasising the importance of developing economically and socially within the constraints established by natural resources. They are less suggestive about the economic and social aspects of sustainability, and are therefore only partial measures of sustainable development. Some of the models discussed here, e.g. material input per unit service and life cycle assessment are particularly focused on material flows in industrial and other processes.

ECOLOGICAL FOOTPRINTS

Derived from the ecological concept of carrying capacity, the ecological footprint is based on the proposition that it is possible to calculate the area of land required in order to support any human activity (Wackernagel and Rees 1996). This area is known as the ecological footprint of the activity. The same concept can be applied to human populations and their aggregate activities. Individuals, families, corporations, cities, states and countries thus have their own ecological footprints.

Ecological footprints have a number of advantages. They are easy to understand, and make comparisons simple. For example, it is easy for a lay person to understand that it requires x hectares of land to sustain the lifestyle of a person in country A but only y hectares to sustain the lifestyle of a person in country B. Such figures have a dramatic and intuitive appeal.

Ecological footprints aid interpretation by making it easy to see certain implications of ecological information. If a particular country has an ecological footprint greater than its own geographic area, then that country is drawing upon the ecological resources of other countries in order to support its own lifestyle (Proops et al. 1999). Similarly, if the ecological footprint for the entire human population is greater than the total area of land available, then human activities in aggregate are unsustainable.

The model suffers from some disadvantages. Prime amongst these is that the ecological footprint is often difficult to calculate. A large number of assumptions are required, and these can be controversial. For example, land area equivalents for energy consumption are found by calculating the area of land needed to grow the equivalent bio-fuels. This assumes that the sun is the only sustainable source of energy and that biofuels are the best means of harvesting it. The ecological footprint model can provide a simple, intuitively appealing measure of the demands that particular activities, organisations, or population groups place on the environment. Provided difficulties with methodologies are not too severe, corporations might consider ecological footprints to summarise environmental performance. Declining ecological footprints coupled with increasing profits would indicate a trend toward sustainability.

SUSTAINABLE PROCESS INDEX

The sustainable process index calculates land area equivalents for goods rather than people or populations. Goods, or production methods, with a smaller land area equivalent are preferred. The methodology is similar to the ecological footprint.

MATERIAL INTENSITY PER SERVICE

Whereas the ecological footprint measures the area of land required to sustain an activity, the material intensity per service (MIPS) calculates to mass of material required to deliver a service (von Weizsacker et al. 1997).

The first step is deciding what service is being delivered. In the service sector, this is self-evident and straightforward. However, MIPS also requires that the unit of service delivered by a good be specified. For example, the unit of service delivered by a car is passenger kilometres. Determining what service is being provided by a good can be the most difficult part of the MIPS calculation.

Once the unit of service is determined, the volume of material required to produce it is calculated using a whole of life cycle approach. That is, every stage of production, use and disposal or recycling is taken into account. This analysis can produce startling results. For example, three tonnes of material is used to produce a gold ring weighing 10 grams and 2.5 tonnes to produce a nine kilogram catalytic converter for a car engine (ibid).

MIPS has many of the same strengths and weaknesses as the ecological footprint model, although difficulties with methodology are less severe. One additional criticism of MIPS is that it does not take into account the different environmental effects of the materials used to produce a unit of service. Displacing one tonne of sand and rock in an arid zone does not have the same environmental impact as discharging a tonne of heavy metals into a biologically rich river system. However, MIPS was never intended to be anything more than a rough indicator of material intensity. Its purpose is to encourage dematerialisation of the economy, i.e. to encourage ways of delivering the same service in a less material intensive way.

MIPS is simplest, and most valuable, when the problem is to compare two ways of providing the same service. For example, the MIPS for pylons to carry overhead 110 kV electric mains wires is eight times higher if the pylons are made of concrete than if they are made of steel (ibid).

LIFE CYCLE ASSESSMENT

Like MIPS, life cycle assessment analyses the total environmental impact of a good or service, including all phases of production, use and disposal. However, the approach is more detailed and sophisticated.

There are four steps in life cycle assessment.

1   Goal definition and scoping, which defines the boundaries of the system under consideration, i.e. the good or service and associated processes, and clarifies the assumptions used in the assessment.

2   Inventory analysis, which identifies and quantifies the material energy inputs and environmental releases over the entire life cycle of the good or service, including raw material acquisition, manufacture, use and disposal.

3   Impact assessment, which examine the environmental impacts of the flows identified in the inventory analysis.

4   Improvement assessment, which looks at opportunities to reduce environmental impacts.

Being more sophisticated than MIPS, life cycle analysis is also more difficult to apply. However, application can be simplified by developing a standard suite of techniques and information to help estimate both the inventory and impacts for a range of goods and services. Extensive databases for these purposes have been developed overseas and are being developed in Australia (Philpott 1996).

ENVIRONMENTAL SPACE

The environmental space model divides the estimated worldwide quantity of each natural resource by the total world population. The resulting per capita natural resource quantities are the 'environmental space' for each person. The environmental space of a nation is obtained simply by multiplying the national population by the per capita environmental space. The quantity of each resource used by a particular nation can then be used to determine whether that country is living within, or exceeding, its environmental space.

There are difficulties with methodology in calculating environmental space, but these are not as severe as those associated with the ecological footprint model. For example, in order to calculate the environmental space associated with greenhouse gas emissions, it is necessary to make some assumptions about the level at which greenhouse gas concentrations should be stabilised. The total annual greenhouse gas emissions that will stabilise concentrations at that level can then be calculated.

BROAD ENVIRONMENTAL MODELS

PRESSURE-STATE-RESPONSE

In the early 1990s the OECD produced the pressure-state-response (PSR) model. In this model, the state of the environment is the quality and quantity of natural resources. Natural environmental variations are considered as part of the 'state'. Human activities are 'pressures', which cause the state of the environment to change. 'Responses' are also human activities, in this case those actions and programmes that are specifically designed to address perceived adverse changes in the state of the environment.

The PSR model is probably the most widely used model for reporting on the state of the environment by governments. In Australia it is used by the Commonwealth, New South Wales, South Australia, Western Australia, Tasmania and Queensland for official state of the environment (SoE) reporting. Most OECD countries also produce SoE reports using the PSR model. It is also suitable for corporate environmental reporting (possibly in combination with one of the management frameworks discussed later in the chapter).

The model is usually fleshed out for reporting purposes by identifying particular issues, themes, or sectors and applying the pressure-state-response approach to each. This provides a 'grid' for presenting information and ensuring all relevant issues are examined fully. For example, Australian SoE reports use seven major themes: biodiversity, land resources, inland waters, estuaries and the sea, the atmosphere, human settlements, and natural and cultural heritage for their reporting framework (SEAC 1996).

A number of minor variations have been made to the pressure-state-response model. For example, some prefer to talk about 'condition' rather than 'state' and style the model as 'condition–pressure–response' to give greater prominence to the natural environment over human activities. The Australian State of the Environment Advisory Committee (SEAC) has announced that in 2001 it will also report on 'implications' as well as condition, pressure and response. The intention is to draw attention more explicitly to those issues that need to be addressed by policy and decision makers. These modifications are designed to enhance effectiveness of the model as a communication tool, and make no difference to the following discussion of strengths and weaknesses.

A strength of the PSR model is that it clearly focuses on how human activities affect the natural environment. This makes it useful for drawing attention to existing and emerging environmental problems that have clear anthropogenic causes. In a comprehensive report, the relative magnitude of these problems can be evaluated and priorities for action drawn up. The model is also designed to draw attention to, and evaluate the effectiveness of, human efforts to remedy environmental problems ('responses'). In practice, this has proved more difficult than identifying the problems themselves. This is partly due to the lack of distinction between 'pressures' and 'responses'. Both are human activities, the only difference being whether the conscious intention is to address perceived environmental problems. Also responses are diffuse and difficult to measure, those measures that are put in place can take many years to have an effect, and many measures that ameliorate environmental problems are made for sound social and economic, rather than environmental, reasons.

One criticism of the PSR model is that it is not a true scientific model. Changes in the environment are driven by non-anthropogenic as well as anthropogenic effects. Another is that the model categorises all human activities as harmful to the environment by citing them as 'pressures'. Critics point out that some human activities benefit the environment, and these are not restricted to activities explicitly intended to remedy environmental damage. Further, even if human activities do cause some harm to the natural environment, some account must be taken of the resulting social and economic benefits. This weakness reflects its design for the purpose of monitoring the environmental effects of human activity rather than as a framework for reporting on sustainable development *per se*.

The recently published *Framework for Public Environmental Reporting: An Australian Approach* (Commonwealth of Australia 2000) recommends that environmental performance indicators be classified as 'operational performance indicators' (which deal with the environmental impacts of the organisations operations), 'management performance indicators' (which deal with the organisation's capacity to manage issues connected with environmental impacts of its operations), and 'environmental condition indicators' (which describe the quality of the natural environment within which the organisation operates). While the nomenclature is different, these groupings correspond to the familiar 'pressure', 'response' and 'state' categories.

DRIVING FORCE-PRESSURE-STATE-IMPACT-RESPONSE

The driving force-pressure-state-impact-response model was developed by the United Nations Environment Programme (van Woerden et al.

1999) as an extension of the pressure-state-response model. This model retains the pressure, state and response elements of the PSR model, and adds two new elements, namely:

- 'Driving forces' — underlying causes of pressures on the environment. These might include demographics, social trends and macroeconomic policy.

- 'Impact' — the effect of changed environmental states upon ecosystems, human health and environmental services.

Because it is similar to the pressure-state-response model, the driving force-pressure-state-impact-response model has similar strengths and weaknesses. An advantage is that the underlying causes of human activities that affect the environment ('driving forces') are explicitly included. The task of relating an analysis of environmental trends and issues to, for example, social or macroeconomic policy is thus simplified.

MODELS BASED ON THE ENVIRONMENT PLUS ECONOMIC AND SOCIAL SERVICES

EXTENDED METABOLISM

Metabolism refers to the flow of energy and materials into and out of a system. It was applied to the study of cities by Wolman (1965), and has since been used in some academic studies of human settlements (for example, Boyden et al. 1981). The metabolism concept can also be used for larger systems than cities; Yencken and Wilkinson (2000) have used it to study energy issues for Australia.

In the case of a city, metabolism involves consuming resource inputs (food, energy, land, water, building materials, other raw materials) and producing waste outputs (emissions to air and water, land fill and other solid wastes). This conversion is effected by the dynamics of the city, which include demographic, social, and economic processes, and are strongly influenced by urban design and infrastructure.

While this model provides a simple and powerful framework in which to consider environmental impacts, it ignores social and economic benefits flowing from the dynamic processes that take place in cities. The extended metabolism model (Newman et al. 1996) was developed to overcome this flaw. It does so by considering a range of 'services' provided by the city.

The extended metabolism model provides a simple and powerful method for deciding whether human settlements are becoming more sustainable. The equations are:

increasing sustainability = falling resource inputs and waste outputs + rising services,

declining sustainability = rising resource inputs and waste outputs + falling services.

This clear, simple method for taking a range of social, economic, and environmental factors into account and determining whether a human settlement is becoming more or less sustainable is the model's greatest strength. However, it should be noted that the extended metabolism model only deals with relative, not absolute, sustainability. There is no rubric for deciding whether the human settlement is sustainable.

As originally applied, the services provided by the city were considered under the heading 'liveability'. Liveability recognises that, in addition to wastes, human settlements produce a range of outputs that contribute to the quality of life of its inhabitants. These can include health services, education and employment, desirable urban environment, community spirit, and access to services. A criticism is that some aspects of liveability are subjectively defined and difficult to measure. For example, what precisely is meant by 'community spirit' or 'desirable urban environment'? This problem can be avoided by defining service outputs in a way that has wider support (although this may lead to a narrower definition).

Although the extended metabolism model was developed in order to report on the sustainability of human settlements, there is no reason why it could not be extended to deal with corporations, government agencies or nations (Newman and Kenworthy 1999). All of these entities apply dynamic processes to material inputs, producing both 'services' and wastes.

For example, a company that produces aluminium might apply the model by listing:

- material inputs as alumina and other raw materials, electricity, land, vehicles and fuel (for transport), and building materials (for plant);

- dynamic processes such as the manufacturing process, marketing, transport and corporate management practices;

- waste outputs as greenhouse gas emissions (from electricity), air pollutants, water pollutants, and any solid wastes going to landfill;

- 'services' as aluminium products, employment, employee satisfaction and profits.

ECOSYSTEM HEALTH

In the last decade scientists have begun to talk increasingly about 'ecosystem health'. Ecosystem health draws an analogy between human health and the health of a landscape or ecosystem. The ecosystem health approach seeks to identify characteristics of ecosystems that make them 'healthy' or 'sick'.

The closely related concept of 'ecosystem integrity' has also been used by a number of researchers and the terms are often used interchangeably, although some claim to find a difference between them (Rapport 1998a). Ecosystem health and/or ecosystem integrity have been defined in a number of ways, and practitioners generally recognise the need to pay attention to context in formulating a definition. Some definitions rely entirely upon natural attributes, generally focusing on ecosystem processes. For example, Mageau et al. (1995) suggest three criteria for ecosystem health:

- vigour (throughput of energy, nutrients, water);
- resilience (the ability to 'bounce back' after stress or disturbance); and
- organisation (the complexity and degree of inter-relationship within the ecosystem).

In a similar vein, Woodley (1993) defines ecosystem integrity as '… a state of ecosystem development that is optimised for its geographic location, including energy input, available water, nutrients and colonisation history'.

Other researchers include criteria relating to human uses of ecosystems in the definition of ecosystem health. To the three criteria identified by Mageau et al. (1995), Rapport (1998b) adds five related to human uses of ecosystems:

- maintenance of ecosystem services (functions such as water purification and production of food from which humans benefit);
- management options (the capacity to support multiple human uses);
- reduced subsidies (the capacity to produce outputs useful to humans with minimal external inputs, and without significant outputs that harm humans);
- minimal or no damage to neighbouring ecosystems;
- minimal or no adverse effects on human health.

Mann (1993) suggests that ecosystem integrity is found in ecosystems that are:

- strong energetic natural ecosystem processes and not severely constrained;
- self-organising in an emerging, evolving way;
- self-defending against invasions by exotic organisms;
- biotic capabilities in reserve to survive and recover from occasional severe crises;
- attractiveness, at least to informed humans; and
- productive of goods and opportunities valued by humans.

There have been a number of efforts to use the characteristics of a healthy ecosystem as criteria against which the success of ecosystem management is evaluated and to develop appropriate indicators within this framework. The system is considered sustainable if the ecosystem remains healthy.

In contrast to some of the models discussed earlier, ecosystem health models have a strong ecocentric rather than anthropocentric focus. Human activities are considered as a perturbation under which ecosystems may or may not continue to function in a healthy way. However, the model becomes less ecocentric and more anthropocentric as criteria relating to human uses of ecosystem services and functions are added. Two Australian examples illustrate the possibilities of using ecosystem health as a model for indicator development. Pankhurst et al. (1997) have developed a set of biological indicators of soil health. These indicators concentrate on biophysical properties, and rely upon a heavily 'natural environment' oriented understanding of ecosystem health. Walker and Reuter (1996) define a healthy catchment as 'one which can recover from perturbations, natural or man-made. It is economically viable and environmentally self-sustaining.'. Three subsets of indicators of catchment health are used: condition indicators, biophysical trend indicators and farm productivity/financial/product quality indicators.

Models based on ecosystem health can offer a strong integrative interdisciplinary scientific basis for considering sustainability. The flip side is that considerable expertise is required to tailor the concept to particular ecosystems and long-term research may be needed. While there has been criticism of the scientific basis of the ecosystem health metaphor (Rapport 1998a), the health metaphor is intuitively appealing to the public, having resonance with everyday concepts of health and sickness, and therefore offers some communication advantages.

While social and economic factors can be incorporated into the model, this is at the cost of scientific objectivity, since value judgements are required. However, even using the most extended set of criteria for ecosystem health, the full range of social and economic factors of interest in sustainable development are not covered. For example, attributes such as literacy and equity of resource distribution are not within the scope of any of the ecosystem health criteria discussed here. The model therefore does not cover all aspects of sustainable development.

SOCIAL, ECONOMIC AND ENVIRONMENT INTERACTIONS

The models in this group are relatively loose 'grids' within which information about environmental, economic and social information is organised in ways that highlight certain relationships between the environment, society and the economy.

DRIVING FORCE-STATE-RESPONSE

Because of its environmental focus, the PSR model is poorly suited to considering broader questions of sustainability. The driving force-state-response model, developed by the Commission for Sustainable Development (United Nations Division for Sustainable Development 1999), is an attempt to adapt the PSR model to monitor sustainability. In this model, 'driving forces' are human activities, processes, and patterns that affect sustainable development. The 'state' is the state of sustainable development. 'Responses' are policies, programmes, and other actions formulated in response to perceived changes in the 'state' of sustainable development.

In the driving force-state-response model, the 'grid' for presenting information is completed using the chapters of Agenda 21 (UNCED 1992), the document developed at the Rio Earth Summit in 1992 as a 'blueprint' for sustainable development. The issues dealt with in the chapters of Agenda 21 are divided into four categories (environmental, social, economic and institutional), and driving force, state and response information developed for each.

The Commission for Sustainable development has developed a set of 130 indicators within the driving force-state-response model, and published methodology sheets for each indicator. The indicators are intended for use at the national level. The European Union has published the results of a pilot study using this methodology (Office for Official Publications of the European Communities 1997).

POPULATION-ENVIRONMENT-PROCESS

The population-environment-process model, developed by Statistics Canada, conceptualises the human population, the non-human environment and the economy as distinct but linked domains (Australian Bureau of Statistics 1996). Attention is drawn to the flow of goods and services from the economy to the population, and resources and services from the natural environment to both the population and the economy. Both the economy and the population 'restructure' (or modify) the natural environment. Pollution is considered as 'leakage' of 'waste stock' from the economy to the natural environment. Bakkes et al. (1994) discuss similar models representing the links between the economy, society and the environment.

GENERAL CRITERIA FOR SUSTAINABILITY

Some models operate by designating universal critieria against which the sustainability of any sector, enterprise or region may be evaluated. The natural step model is an example (refer Chapter 2).

THE NATURAL STEP

This model was developed in 1989 by Karl-Henrik Robert and a panel of Swedish scientists, and is fostered by an organisation with branches in the United States, Canada, the United Kingdom, Sweden (among others) and Australia (Natural Step 2000). It specifies four general criteria for sustainability, known in the Natural Step framework as system conditions:

- In order for a society to be sustainable, nature's functions and diversity are not systematically subject to increasing concentrations of substances extracted from the Earth's crust.

- In order for a society to be sustainable, nature's functions are not systematically subject to increasing concentrations of substances produced by society.

- In order for a society to be sustainable, nature's functions are not systematically impoverished by physical displacement, over-harvesting or other forms of ecosystem manipulation.

- In a sustainable society resources are used fairly and efficiently in order to meet basic human needs globally.

In addition to the four system conditions, the natural step encourages 'backcasting' (thinking about current activities and short-term objectives in light of long-term sustainability goals), 'systems thinking' and a 'step by step' approach to achieving sustainability.

A number of corporations and local governments are using the Natural Step as a framework for developing systems for monitoring and reporting on sustainability. The Natural Step has the advantage of a simple and flexible framework, combined with a pragmatic stepwise approach to achieving sustainability goals. This has a clear appeal to corporations. On the other hand, the Natural Step makes a particular interpretation of sustainability that some organisations may find unacceptable.

SPECIFIC CRITERIA FOR SUSTAINABILITY

The models in this group operate by specifying criteria that must be satisfied in order for a particular activity, sector, city or country to be sustainable. This section includes examples of specific criteria designed for a nation (Sustainable Britain), a sector (Criteria and Indicators for the Conservation and Management of Temperate and Boreal Forests), and a region (Local Agenda 21).

CRITERIA AND INDICATORS FOR THE CONSERVATION AND SUSTAINABLE MANAGEMENT OF TEMPERATE AND BOREAL FORESTS

The criteria and indicators for conservation and sustainable management of temperate and boreal forests were agreed in February 1995 by the governments of Argentina, Australia, Canada, Chile, China, Japan, Republic of Korea, Mexico, New Zealand, Russia, the United States of America, and Uruguay within whose borders 90 per cent of the world's temperate and boreal forests lie (Montreal Process Implementation Group for Australia 1997). The Montreal Process criteria and indicators are an excellent example of a model based on criteria of sustainability for a specific sector (i.e. management of temperate and boreal forests).

The Montreal Process criteria are:

- conservation of biological diversity;

- maintenance of productive capacity of forest ecosystems;

- maintenance of forest ecosystem health and vitality;

- conservation and maintenance of soil and water resources;

- maintenance of forest contribution to global carbon cycles;

- maintenance and enhancement of long-sterm multiple socio-economic benefits to meet the needs of societies.

The Montreal Process criteria are, essentially, an internationally agreed set of conditions for the sustainable management of temperate and boreal forests. If the criteria are satisfied, forest management is deemed sustainable. In addition to the criteria, the Montreal Process has agreed a set of 67 indicators, grouped under the various criteria, to provide measures of whether forest management is meeting the criteria. In 1997 Montreal Process countries presented first approximation reports based on the criteria and indicators. The object of the first approximation reports was to test the feasibility of the model, rather than to decide whether forest management is sustainable.

SUSTAINABILITY COUNTS

In November 1998 the United Kingdom Department of the Environment, Transport and the Regions released a discussion paper entitled *Sustainability Counts* (DETR 1998) that contained a proposed set of 13 'headline' indicators of sustainable development for the United Kingdom. The headline indicators were based on four broad objectives:

- maintenance of high and stable levels of economic growth and employment

- social progress that recognises the needs of everyone

- effective protection of the environment

- prudent use of natural resources

These broad objectives, which had been the subject of public consultation in the preceding 12 months, constituted a proposed set of criteria for determining whether development in the United Kingdom is sustainable. In the words of *Sustainability Counts*: 'achieving sustainable development means addressing all of these objectives equally, both for present and future generations'.

LOCAL AGENDA 21

Local Agenda 21 is an initiative endorsed by the United Nations and its member states to encourage local governments to use the framework of Agenda 21 as a basis for creating local strategies for sustainable development. It is a loose framework that draws attention to key issues and encourages local communities to work consultatively to produce their own priorities and strategies for sustainable development.

The Local Agenda 21 framework considers economic, social, environmental and institutional factors in an integrated way. Indicators of sustainable development are to be created within this framework and tied to the locally determined strategies and priorities.

A 1996 survey of local governments in Australia (ALGA 1996) showed that 121 councils were actively developing local sustainability strategies (or something similar), 43 of them using the Local Agenda 21 framework. Examples include the City of Adelaide (City of Adelaide 1996) and the City of South Sydney (South Sydney City Council 1995).

DISCUSSION

Models based on specific sustainability criteria are suitable for dealing with complex issues where it is important to take a full range of social, economic and environmental factors into account.

Specific criteria models are useful where multiple interests are involved and arguments can be made that, within certain boundaries defined largely by ecological and social realities, a range of development paths are sustainable. Widespread consultation can be used to build consensus around a set of sustainability criteria for the sector, region or activity of interest. None of the other models surveyed in this chapter are capable of properly taking the complex range of interests and issues into account. While governments have the resources necessary to consult widely where issues and public interest dictate, corporations may be less willing to undertake such major exercises. For this reason, among others, they may find other models more suitable.

A further potential disadvantage is that consensus may be achieved at the cost of meaning. Sustainability criteria may be too vague for a precise interpretation, allowing interest groups to assign to them whatever meaning they like. The criteria would then be useless for deciding whether development is sustainable, or at least becoming less unsustainable. A related danger is that the criteria will become badly compromised, ignoring scientific evidence about the limits within which development must occur in order to be sustainable. It is therefore important that discussions about sustainability criteria be well informed.

It is often advantageous to make comparisons between cities, countries, sectors and corporations. For example, comparing the sustainability of development may reveal where additional efforts are most needed to boost performance. Using specific criteria can both enhance and detract from comparability.

Comparability can be enhanced if the regions, countries, sectors, or corporations being compared are using the same set of criteria. An advantage of approaches based on criteria is that they allow indicators and other analytical tools to vary from entity to entity so that local conditions can be taken into account. On the other hand, comparability is reduced if entities are using completely different criteria of sustainability. In extreme cases, these criteria may be incommensurate.

In Australia models based on specific criteria have been used successfully in a number of sectors. The Standing Committee on Agriculture and Resource Management (SCARM 1998) has used five criteria as a basis for developing and reporting on a set of sustainability indicators for agriculture in Australia:

- long-term viability and resilience of farm economies;
- quality of farm management skills;
- socio-economic viability of rural communities;
- minimisation of off-site environmental impacts; and
- enhancement of the resource base.

The Bureau of Rural Sciences (Chesson and Clayton 1998) has developed a set of criteria for the sustainability of fisheries. The framework considers effects on humans (food, employment, income lifestyle) and on the environment (primary commercial species, non-target species, and other aspects such as the marine landscape, water quality and the movement of organisms).

MANAGEMENT PRINCIPLES

Management principles and frameworks can be used to organise and interpret information about sustainable development. Such an approach has two strong advantages. Firstly, it is not necessary to develop or adapt new models although existing management frameworks will have to be modified. Secondly, there is already a clear link between information and decision making. Here two approaches are considered. One based on performance measurement and the other on continuous improvement or adaptive management.

PERFORMANCE MEASUREMENT

Many organisations routinely measure performance against stated goals. The performance measurement model adopted by the Australian Department of Finance (1994) is a representative approach. This approach incorporates the following elements:

- goals: what the organisation seeks to achieve;
- inputs: the programmes the organisation has in place and the resources committed to them;
- outputs: the goods and services that the organisation produces directly (these may be motor vehicles in the case of a car manufacturer, or social security payments in the case of a government department);
- outcomes: these are the effects that the organisation's outputs have (in the case of a government department responsible for social security pay-

ments, the outcome may be fewer people living in poverty; for a car manufacturer it may be profits).

Two key measures of performance may be derived from this model: effectiveness, which is the degree to which outcomes accord with goals, and efficiency, which is the ratio of inputs to outputs.

This model (and similar models) can easily be adapted to monitoring sustainability. The key is to incorporate sustainable development into organisational goals. A clearly articulated picture of what sustainability means for the organisation is essential. This will require careful thought, and almost certainly involve changing some aspects of operations.

Consider the hypothetical example of a motor vehicle manufacturer. For such an organisation, goals might be expanded from the traditional corporate goal of profit to include reduced air pollution, increased safety, greater worker satisfaction, and reduced solid waste from motor vehicles and motor vehicle production. Much of this broadening of vision has already taken place in many major manufacturers. Adjustments to goals will be carried into changes to inputs. For instance, the type of vehicles produced and the way they are marketed and manufactured. This will lead, in turn, to different outputs and outcomes. In order to inform management about sustainability, it will be necessary to find ways of measuring outcomes that are not related to profit. Measures such as the level of air pollution in major cities, number of cars going to landfill, number of deaths and injuries from motor traffic accidents and surveys of employee satisfaction may be relevant.

Perhaps the most difficult aspect of applying this model is finding ways to measure outcomes and relate them to outputs. While outputs are entirely within the organisation's control, outcomes nearly always depend on a range of outside influences. In the case of our hypothetical car manufacturer, air pollution levels, the number of cars going to landfill and the number of deaths and injuries from motor traffic accidents will depend on consumer behaviour and other industries as much as on the nature of the cars produced.

ADAPTIVE MANAGEMENT AND CONTINUOUS IMPROVEMENT

Adaptive management recognises that natural systems are dynamic, subject to multiple influences, partially understood and subject to unpredictable changes. Iterative changes to management approaches, with constant feedback and re-evaluation, is the hallmark of adaptive management.

One approach to adaptive management adopted by the Australian Local Government Association (Thorman and Heath 1997) recognises five levels of management planning: vision, issues, objectives, targets and actions. These different levels of planning need to be reviewed at different intervals. One year is long enough to decide whether many actions are effective, but a quarter of a century may be required before the vision changes. The nature of adaptive management is such that continuous monitoring is required. Monitoring systems must be put in place that will allow managers to decide whether visions are being achieved, issues are being adequately addressed, new issues are emerging, objectives and targets are being achieved, and actions are effective.

Continuous improvement is closely related to adaptive management. Continuous improvement seeks constant, goal-oriented, incremental changes in management and operations. The environmental management system developed by the International Organization for Standardisation (ISO) is an example of a continuous improvement model. The Australian and New Zealand Standard ISO 14001 is derived from this model (Standards Australia 1996). The standard requires organisations to develop environmental policies and plans, implement them, monitor their effectiveness, and regularly review and modify them. The International Organization for Standardisation has prepared further guidance on environmental performance evaluation in its standards ISO/DIS 14031, and 14032 (ISO Secretariat 1998).

Adaptive management and continuous improvement involve intensive performance evaluation. Accordingly, the key to incorporating sustainability into the monitoring and evaluation systems is the same as in the performance measurement model. Sustainability must be part of the vision or plan, and must be allowed to flow into issues, objectives, targets and actions in a meaningful way.

SELECTING A SUITABLE MODEL

There is no objective way to decide which is the 'best' model for monitoring and reporting on sustainability. All models have strengths and weaknesses and the preferred model varies from situation to situation. A number of factors ultimately influence the selection of a model.

INFORMATION REQUIREMENTS

Some models may be unsuitable if the specific information or methodology required is unavailable, uncertain or difficult to obtain. For example, the ecological footprint model requires a methodology for converting all resources into equivalent areas of land. While there are methodologies available for doing this for some resources, it is more difficult for others.

AUDIENCE

The audience for monitoring and reporting needs to be considered. Models that require a greater level of technical expertise and sophistication may be suitable if the outputs are to be used by technical specialists and natural resource managers. However for decision makers, policy makers and the public a more graphic model may be suitable. For example, the ecological footprint model can help to bring important issues to the attention of the public and policy makers, but may not be particularly useful for detailed analysis of possible solutions.

SCOPE AND SCALE

Another important question to consider when selecting a model is whether it is to be used for monitoring and reporting on sustainability *per se* or the environment. Some models are better suited to environmental monitoring and reporting rather than the broader considerations of sustainability. Broader sustainability issues may be less important for monitoring and reporting at smaller geographic scales or for corporate entities. At these scales entities may report on how they are contributing toward achieving global goals such as ozone depletion or greenhouse gas emissions but they cannot realistically report on whether these goals are being met.

Models that can answer the question 'is development sustainable' are more suited for monitoring and reporting at national and global scales, while models that answer questions such as 'is this activity/region/settlement becoming more or less sustainable?' are better suited for reporting at regional and local scales.

VALUES

While all models include assumptions, some are more value-laden than others. In selecting a model it is important to examine the underlying assumptions carefully and be sure that the explicit and implied values are acceptable. Many models have been designed on the basis of a particular understanding of sustainable development. These models tend to be those that are most suggestive of particular interpretations of information.

NATURE OF ISSUES OR ENTERPRISE

For complex or poorly understood issues flexible models that require a high degree of expert input may be suitable. Ecosystem health and models based on specific criteria are examples. Where issues are controversial, two approaches may be taken. One is to consult widely with interested parties and technical experts to develop a consensus model for monitoring and reporting. The alternative is to use a relatively neutral framework that suggests little about the meaning of the information organised within it. The latter approach avoids controversy while the model is being developed, but offers less guidance about whether development is sustainable.

PRECEDENTS AND NEED FOR COMPARABILITY

Where comparisons between regions, time periods or enterprises are important, it is necessary to consider the models that have been used previously. For example, if a corporation seeks a commercial advantage by showing that its activities are more sustainable than those of its competitors (or making them so), it may be important to use the same model as competitors to monitor and report on sustainability.

MIXING MODELS

In some situations it may be desirable to mix models to derive the most suitable approach. For example, corporations may adopt a continuous improvement approach to managing the environmental impacts of their operations, while using a PSR model, or its functional equivalent, to organise environmental information. Another example is the use of 'forest ecosystem health and vitality' as a criterion in the Montreal Process. The ecosystem health approach is thus nested within criteria for the sustainable management of forests.

MULTIPLE MODELS

An entity need not be restricted to using one model. It is possible to use the same data to support two or more models, which can be used to inform different stakeholders or decision makers. For example, in Australia vegetation cover is of interest to both the Montreal Process and SoE reporting (Saunders et al. 1998). The Montreal Process uses a 'specific criteria' model and is interested in information about the areal extent and structure of forests. SoE reporting uses a PSR approach and is interested in information about the areal extent and structure of all vegetation types.

TRENDS AND FUTURE DIRECTIONS

Sustainability is a complex issue, and the most pertinent questions vary with scale and the nature of the entity, e.g. community, manufacturer, financial institution, Federal Government, and industry sector. It is therefore reasonable to expect that a range of models will be used. Accordingly, trends are difficult to identify, and vary with the nature of the entity.

CORPORATE

The last decade has seen an increasing number of large and medium-sized corporations prepare public environment reports, which present information about how the corporation's activities affect the environment. Guidelines for such reports have been prepared by the United Nations Environment Programme (UNEP 1996) and Environment Australia (Commonwealth of Australia 2000), among others.

This activity is motivated by a desire to make the environmental impacts of corporate operations transparent, and the perceived business advantages of doing so. There are also some statutory requirements for environmental reporting in many countries, including Australia (Commonwealth of Australia 2000).

Some corporations have also begun to experiment with 'sustainability reporting' (for example, Shell 1998), which goes beyond public environmental reporting to include social effects of corporations' activities. The Coalition for Environmentally Responsible Economies (Global Reporting Initiative 1999) has published a draft framework for sustainability reporting, but acknowledges that more work is required on its conceptual basis.

Models used in public environment reports are not sophisticated, and probably do not need to be. As already noted, reports tend to be based on a performance management approach, with a PSR type model often used to organise information. As the trend toward sustainability reporting accelerates, more sophisticated models will be required. Models, such as the extended metabolism model and the natural step, have strong potential for corporate sustainability reporting. Some corporations may also be attracted to models derived from economic approaches, such as those discussed in Chapter 4 of this book.

GOVERNMENT

Most OECD countries produce SoE reports using a PSR model. In Australia, most states and territories also produce reports using this model. This model has now been in common use for almost a decade and shows signs of being well established.

Models based on specific criteria have been used to develop information about the sustainability of some sectors of Australian industry. The Government of the United Kingdom has used this approach to develop a set of 'headline' sustainability indicators. Approaches based on models derived from economics (such as satellite accounts — see Chapter 4) are also being actively developed in many countries, including Australia.

Where nations choose to report on sustainable development (whether at the national or sector levels), developing specific criteria is likely to prove the most successful approach. It has the advantage of allowing for variations in the circumstances and aspirations of different countries, and offers better guidance on whether development is sustainable than do models such as driving force-state-response.

4
ECONOMIC MODELS
JOHN HATCH

INTRODUCTION — ECONOMICS AND ITS HUBRIS

Economics is a relatively young science. Of course some people will query whether it is a science in any real sense at all, preferring to see it as a rather narrow, amoral religious creed. Whatever one's view it clearly has been very successful especially in the 20th century and now the 21st, when its doctrines seem to intrude into every facet of life. This has been the more so in recent years when the collapse of the 'Evil Empire' and the relative success of Western capitalism have seemed to endorse conventional neo-classical economics, not just as a description and analysis of a system, but almost a description and analysis of *the only possible* system. Can this would-be science help us to achieve sustainability? Can it really encompass sustainability or is it likely to be embarrassed by this ill defined but almost universal objective? After all, theories and models, particularly in economics are driven by objectives, human objectives, thus we must consider whether our current economic 'theories' are compatible with this new objective.

Descriptions of economics, at least those written by economists, tend to talk about choice, allocation of resources and the like and they quite often also mention words like 'wise' or 'best'. Thus economics not only often claims to be a science but also in some senses a moral and indeed a highly perfected science. There is a unanimity in modern economics about method and purpose that has perplexed some, but is a great source of strength for the discipline and most practitioners. Thomas Kuhn (1970) captured this in his brief and only reference to economics, when he said 'It may, for example, be significant that economists argue less about whether their field is a science than do practitioners of some other fields in social science. Is this because economists know what science is? *Or is it rather economics about which they agree!'* (my emphasis). He is probably emphasising the disconcerting, almost frightening uniformity (coherence) of economic thought that is codified in the dominant paradigm — neo-classical economics. This complacency and certitude may have led Geoff Harcourt to coin the term 'The Social Science Imperialists' (Harcourt 1982).

Modern economics has at times given the impression that its methods can be used to explain all human behaviour and we have had, courtesy of Gary Becker and others, economic models to explain marriage and divorce, suicide and just about every other human action. Can one doubt therefore that such models can elucidate sustainability? As Hamilton (1994) says, 'The idea that we can conquer the world by analysing it is precisely the attitude of economics', and for many economists the economic model of individual choice in the market place is the starting point. Almost all problems of human organisation can be solved within this framework and it is this confident and holistic model that we must evaluate, even though as we shall see, there are other views even among economists.

THE FUNDAMENTALS OF THE NEO-CLASSICAL ECONOMIC MODEL

Modern economics as a coherent discipline is usually seen as having its origins in Adam Smith's great work, *The Wealth of Nations*, published in 1776. In this he outlined a very eclectic and wide-ranging social and moral system that of course included the so-called invisible hand of the market, the latter based on individualism, competition and freedom. Such a system is essentially democratic and highly attractive and in its purest form is highly inclusive and disinterested, and leads to very desirable social outcomes. It has emerged as we know as modern capitalism, but is it sustainable, or more practically, can it encompass notions of sustainability as expressed now?

What is the very essence of neo-classical economic theory? It is based on freedom of choice for the well-informed individual. Emphasis is on the individual and this ultimately translates into the notion of rational economic man (and woman). The whole system is assumed to be driven by such individuals who attempt to maximise their own personal satisfaction through pursuit of individual preferences. Of course at times, it is recognised that they may act as social groups — most obviously the concept of the family or household. Despite ideas to the contrary, there is nothing inherent in economics that precludes altruism as it is generally understood, so that it is misleading to argue that the economic model is entirely based on selfishness. It certainly assumes self-interested behaviour, but in such a way that this is allowed to include almost any normal (rational) human behaviour.

Indeed, as a critic, Hamilton (1994) has recently said, 'Fortunately for economics the basic ideal of rational economic man can be moulded so that it incorporates some additional aspects of the world that influence economic decision making ...'. In addition, it is one of the triumphs of Adam Smith's conception that individual actions, driven primarily by individual self-interest freely conceived, will under certain circumstances lead to outcomes that are socially and even ethically desirable. The latter point is contentious since the market model is basically at best amoral; some think it immoral, often by default. Of course the 'certain circumstances' are not unrestrictive. Individuals must have genuine freedom of choice in consumption and production activities. They must have good information and in general no one should have significantly better information than others. In general this means that there needs to be large numbers of economic actors in all activities so that no individual or small group has significant power — the threat of monopoly! Anyone who has had even a passing brush with modern economics will recognise these conditions as the basis of the competitive market model and further know that it theoretically delivers some remarkable outcomes, most of which have admirable features. Attractive, but not perfect!

In practice of course the world does not always conform to the requirements of even this wide-ranging and inclusive model. It doesn't always work well at its boundaries. Economists have for long dwelt on the *technical* reasons for what they call market failure, the most obvious being the presence of market power, more commonly known as monopoly. That is seen as evil and as a technical failure. More important than this, but more intractable are the problems of externalities (spillovers) and those of information, and even more important for our purposes I believe are the boundary problems of modern economic theory. The things that go wrong with economics and the market model in particular, when it is pushed into areas for which is was not really designed, indeed areas that were not really envisaged while the model was evolving. The environment in the broadest sense is one of these areas but by no means the only one.

I have looked at these boundary problems briefly (Hatch 1995) and intend to do so in a book that is in preparation with the tentative title, *Too Much of a Good Thing*. The 'Good Thing' is the market, which as we know is seen as a sort of universal solution or if you like, an almost perfect system of allocation of resources. However, before we look at these very broad criticisms of the market we must consider its failings within its own context and how it has attempted to deal with these. Many of the failings, and the responses to them, are germane to questions of sustainability and other related environmental issues.

We have already discussed the very broad basis of market capitalism in individualism, freedom of choice and atomism. In this way it perhaps loosely mirrors physics, being built on smaller concepts that then aggregate, often *rather simply or even simplistically* into bigger structures. What are the details of this model and can they encompass sustainability and related ideas? These questions can be discussed in terms of dimensions — time, ambit and species — let us take them in reverse order.

SPECIES

While this issue is not absolutely germane to the issue of sustainability, it does affect the whole nature of economic theory and how it is related to such issues. Economic theory in the guise of neo-classical economics is absolutely and unashamedly anthropocentric. It is about the human species; it is logically and structurally bedded in human aspirations and preferences. There is no direct place for the preferences of apes let alone the myriad of other less advanced life forms. In a famous exchange, Beckerman (1972) cut through the philosophical mire when he said, 'It is a basic proposition of welfare economics that commodities cannot enter directly into any social welfare function but only their contribution to somebody's utility function'. Sustainability in some sense must be an anthropocentric concept.

This was a very honest if brutal response to the criticism that economics is narrow and anthropocentric. Of course, the import of it is that we can in fact incorporate the 'interests' of non-human life, but only through the preferences of sympathetic human beings, and we now know that such human beings are becoming more numerous and more influential and noisy. But they are still *only* human beings and cannot enunciate directly the preferences of whales and apes. We can of course subsume these preferences in a crude fashion by invoking intrinsic value, but it is hopelessly vague and unquantifiable and inevitably loses out to market values in conventional analysis. In any case it is itself a human concept and does not in any sense meet the fundamental criticism. The need to incorporate non-human values and preferences becomes more important as one becomes more aware of the intellectual capacities of higher life forms.

We now realise that few things absolutely separate us from the animal kingdom. They arguably have language, use tools, and indulge in forms of social and altruistic behaviour not altogether unlike us. At the margin, that concept so beloved of economists, arguably they are in the shape of the great apes extraordinarily close to us in their genetic make up. Further, modern transgenic work has increasingly shown that even non-simians are variously very close to us in a wide range of body functions. We do medical research and grow transplant organs in rats, pigs and sheep, and the broad justification is that these creatures are variously like us. Philosophically, this sea change is expressed by Singer (1975, 1985), where the fundamental concept of speciesism is

developed and discussed. This concept that places man firmly in a continuum with the animal kingdom really does question the validity of a
model that wholly separates man from nature.

Singer crucially distinguishes 'sentient' from 'non-sentient' beings,
others, as we have seen later, have invoked the more radical idea of
intrinsic values. As we know sustainability is a holistic concept and
therefore may founder on this weakness of the economic model. We
will return to this question later because it cannot be avoided by cant
and sophistry. Beckerman to his credit did not attempt to avoid it, but
like many other economists he fully 'confessed' the weakness, without
admitting its relevance. Twenty-five years later we cannot deem it
wholly irrelevant to sustainability. Indeed, given that sustainability is
such a holistic concept that brings in the whole natural world, we must
confront the issue. Man is not alone on this planet and even economic man must recognise his reliance on the richness of biodiversity not
only now, but into a sustainable future.

AMBIT

Another problem, which is inherent to economic theory, is that in the
interests of simplification, it treats markets as partial or separated, at
least in the basic analysis. Thus the price of apples is assumed at least
initially, to be quite separate from the price of pears. This is achieved
by assuming that while we analyse the price of apples, the price of pears
and indeed all other goods remain constant. Not only do other prices
remain constant, but so do a whole gamut of things, incomes and
tastes. This of course is a familiar methodology not confined to economics and indeed it is very important in making the analysis simple
enough to be manageable. In economics we call it the *ceteris paribus*
assumption, in other words 'other things remain equal or the same'.
The problem is that while it is defensible methodogically — indeed it
is essential — it does tend eventually by osmosis, to engender a nar-
rowly complacent attitude of mind that may be inimical to thinking
about big concepts like 'sustainability'. True, economists do analyse
what they call general equilibrium that allows that all things might
influence all others, but this is the realm of high theory, not everyday
use of economics. Partial analysis is very powerful but it is also very
limited. We might note that this issue is very different from that of spatial analysis that is often covered by economic geographers.

A very special problem relates to the outer boundaries of market
analysis. In analysing the price of apples we initially exclude the price
of pears as a factor, but we allow it in again with reasonable comfort,
it just makes the analysis a little more complicated. However, when we
get to issues such as whether spraying apple crops with insecticide
destroys wildlife we invoke the concept of externalities or spillovers.
These are matters that we see as inherently outside the main game —

the market — and have proved to be difficult to incorporate fully. They are another true boundary to economics in the sense that the only solution that appeals to economists is to somehow bring them into the fold — to internalise them — in other words all such effects must be priced or costed and thus become commodities, exchangeable and quantifiable. Environmental things are nearly all by their very nature externalities and much of the debate about the use of economics hinges on the issue of how to deal with them. Do we internalise them or do we attempt to rewrite economics or do we just give up and leave them out? This issue is the nub of environmental economics issue and therefore germane to sustainability. This idea was famously picked up in Hardin's paper *Tragedy of the Commons* (Hardin 1968), which analyses the nature of so-called common goods. Things owned by no one in particular and thus not subject to the discipline of market forces.

TIME

Clearly the concept of sustainability, however defined, must include time. In Chapters 1 and 2 of this book, authors have wrestled with the problem of defining the concept but it must include some notion of continuation, having a future path and the like. Many of us lean towards the Brundtland Report *Our Common Future* (1987) type of definition and indeed in broad terms this has become a consensus definition accepted in broad terms by environmentalists and economists.

Time has however always been something of a problem for economics. First and foremost, time allows almost everything to change and thus massively complicates an already very complicated world. Secondly, if one is concerned about issues like fairness, justice and the like, then time hugely increases these as well. In longer periods of time we are asked to evaluate such issues when they include non-contemporaneous groups of humans, different generations and of course in the even longer period, we need to evaluate the interests of non-overlapping groups, and this raises problems for traditional concepts of democracy. The preferences of future generations of humans may be almost as inaccessible as those of present generations of non-human animals. Needless to say economists have devised methods of incorporating these future generations into the decision calculus, but none of them are wholly satisfactory, as we shall see later.

Basic economic analysis is very present-orientated and to achieve this it tends to employ rather a restricted methodology usually known as 'comparative statics'. This requires one to ignore time or assume it away in the sense that the model always assumes movement towards a fixed resting-place, an equilibrium. It is not denied that the world constantly changes and therefore that these equilibria are in no sense permanent, but it is the attainment of an equilibrium that is the end point. We can then look at factors that define a new equilibrium. Thus the analysis is in no real sense dynamic or continuous, but proceeds by a series of punctuated changes. Again, this view of the world, which emphasises short-term adjustments, is not conducive to long-term views and analysis. More sophisticated economic analysis does incorporate true dynamics and there is a whole branch of economics that looks at long-term 'adjustment', but calls it growth.

At this point it is important to emphasise that economics in general sees growth as a primary objective and does not recognise any absolute or binding constraints. In Chapter 1, there is extensive discussion of this growth paradigm and its consequences. Because growth — increase — is seen as absolutely central to the analysis, it is not surprising that so much attention is paid to efficiency. Of course, economists define efficiency in a rather special way, basically using resources to produce as much as possible. One of the casualties of this is that distribution — who gets the goodies — tends to take second place. In fact, it does not fall out naturally from the analysis and the stock answer is that distribution is not a 'scientific' question, but one solely of values. This is of course true. If you were to ask a hundred people, let alone a hundred economists, how we should distribute the multitude of goods and services generated by a modern economy, amongst its population, they would probably all come up with different answers. We get different answers, not only because of self-interest or even malice, but rather because of genuine differences in values and the corollary that economics prefers to discuss efficiency.

Of course this preference is thrown into stark relief when one considers questions of sustainability, since the nub of the question is, 'What about the future?' and by implication how do we bring future generations, mostly yet unborn, into the decision-making process. One simple answer is to generally assume that they will be very similar to present generations, another is to ignore them except to allow our own preferences on their behalf and yet another is to try by various means to second-guess their needs and aspirations. None of these is entirely satisfactory. Perhaps this is why even environmental economists at times seem to fight shy of the issue.

Common (1995) in an otherwise splendid and compendious book, relegates the issue of time to a few pages and sees it primarily as a technical problem for cost-benefit analysis (CBA). This is all right as far as it goes but surely the issue warrants broader and deeper discussion. Certainly, he canvasses the ideas that in CBA it has been argued that long-term costs and benefits should be discounted at zero and that a compromise may be to discount them at a lower rate than other impacts, but who is to say at what rate? The notion of discounting relates to the idea that in general people value nice things more highly the earlier they get them. The out for him is to say that if these long-term impacts are seen as incommensurable, then CBA should not be used. He then implies that in fact these impacts can be seen as in '... the consumer sovereignty framework' and as such as amenable to conventional CBA analysis. It all seems rather defensive of what he calls '... standard economics' and as such not really awfully useful unless your prime objective is to defend economics as opposed to reconciling a broader concept of the subject with the immense complexity of the time issue. In all fairness to Common, he does end by saying that in 'standard economics' it is dangerous to mix efficiency and equity. He goes on to point out that the equity issues of time tend to become important when we have large projects — meaning projects that do affect the distribution of total consumption over time.

As is almost always true in the social sciences there is a large grey area and many projects (activities) are neither clearly large nor small. This is very awkward for economics, but it is the way the world is. Thus small projects can use CBA, large projects cannot and middle-sized projects in all dimensions tend to be in limbo. The problem is that a lot of projects are middle-sized and an increasing number of our impacts are definitely large in all senses. These activities, destruction of biodiversity, Enhanced Greenhouse Effects (EGEs) and the like, present equity problems that spill over into the distant future and that may be a continuing function of our present style of activity. As such we do seem to have no really satisfactory way of dealing with time so long as we continue to entertain ideas of democracy.

WIDER ISSUES OF MORALITY AND FUNGIBILITY

A rather different approach is adopted by Hamilton (1994) when he says, 'To the extent that decisions about the natural environment involve moral choices — and surely the condition in which we leave the natural world for future generations is one of the weightiest moral choices — discounting can only cloud the issue'. He goes on to say that the future state of the Earth for many people is not a 'commensurable' good, at least not commensurable with other goods and that, 'the idea of attaching prices to the environment or aspects of it appears strange, indeed abhorrent, even if choices do have to made between preserving the environment and exploiting it for financial gain'. This of course calls into question the whole idea of exchange and markets as the only organising system. The notion that everything is fungible is at the root of this. To be fungible is to be exchangeable or convertible. To make something fungible often diminishes it and certainly has the potential to change it. It also means that naturally nothing is absolute or is an end in itself. It invites one to consider *everything* as a means to other ends and in the extreme to any and all other ends. For this reason if for no other it is 'offensive' to many people. When one considers in addition that economics at best is inherently amoral, and perhaps even immoral, this distaste becomes, I believe, more understandable.

Hamilton, amongst others, distinguishes clearly between ethics and economics and he says, 'The tragic error of economics, and of ourselves when we allow ourselves to be possessed by *homo-economicus*, is to believe that the selfish calculator is *all* that we are' (his emphasis). Later on he is even more trenchant when he says:

'It is not a question, then, of which discount rate is appropriate to the environment. Discounting is simply inappropriate for activities or development that involve a significant environmental component. Unlike the commodities of traditional economics, which people prefer now rather than in the future, the environment cannot be thought of as a commodity for which there is a money value. People do not value the environment less in the future than they do now' ...!

I have quoted extensively from Hamilton not because he resolves the economists' view of time but because I believe that he realistically confronts it without shirking. The environment is not a set of fungible commodities and it almost certainly cannot be usefully squeezed into that mould. Other writers in this book will no doubt attempt to squeeze most of it in but I believe that this is essentially futile and indeed does a disservice to economics and to the environment. Of course, at this point one might despair of economics and economic models as having any role in our attempt to elucidate sustainability. However there are other ways of considering these issues and these will be discussed later.

The theme so far has been of the failings of traditional or standard economics based on its treatment of species, ambit and time. On all three fronts it is perhaps wanting and may be irredeemable. However, it would be quite unfair not to discuss the valiant efforts of environmental economists to build a sub-discipline that addresses all of the these issues and in doing so bridges some of the gaps.

THE ROLE OF SUBSTITUTION

One of the most powerful concepts in economics is substitution and it is indeed a rich concept often grossly undervalued by non-economists. It is often seen as an almost one-dimensional concept, where one item, a good or input is simply replaced by another often at one point in time. In practice, substitution is an on-going process in a market economy and of course it is in response to both the technical possibilities and to the flux of relative prices, and the complex interaction of these two.

Consider the so-called oil crisis of the 1970s, and the responses to it. They richly demonstrate the substitution process and its ability to deal with change and in that case a rather sudden and artificial shortage. We now know, as some economists did then, that the world had not suddenly run out of oil. What had happened was that a dominant group of oil producers, OPEC, had flexed their collective market muscle and orchestrated an oil shortage. Consider the responses that occurred, many of them within days, but some in weeks and months. As the price of oil rose people bought less petrol at the pump, lower speed limits were introduced and in the USA this produced large fuel savings. In months people bought smaller and more fuel-efficient vehicles and switched to other forms of power in production processes and heating. In years, projects such as the US Supersonic Transport were shelved and the relatively fuel efficient Boeing 747 became the dominant form of aircraft. As the high price persisted, oil from coal became feasible and with the help of political events became a reality in South Africa. Oil search became much more profitable and undersea oil sources began to be exploited, as did various more marginal sources on land. All these processes are part of the overall substitution process initiated by relatively high oil prices. As we now know they collectively 'solved' the shortage so that the real price of oil remained low for another 20 years until it rose again recently in response to another OPEC initiated shortage. My purpose is not to discuss the price of oil, nor to argue about whether it is likely to run out soon. Rather it is to show that substitution must be taken seriously as a factor in thinking about sustainability. As a theoretical concept substitution is central to most conventional economic approaches to sustainability. Other less dramatic but more persistent types of substitution involve recycling, substitution of one factor input for another, and the role of technical progress in changing the various options in consumption, production and waste assimilation.

Using this universal and powerful concept economists have developed the so-called Hartwick rule, (Hartwick 1977 and more accessibly Hartwick and Oliwiler 1998). This is very much the *economists'* approach to sustainability since it starts by asking the very narrow *anthropocentric* question, 'What is the maximum sustainable rate of consumption' (per capita) (Common 1995)? In the presence of a finite stock of essential non-renewable resources, the commonsense answer must be zero but this is clearly unacceptable. Enter substitution, but in a rather prescribed and specialised way. In relatively plain English, the Hartwick rule assumes that with efficient extraction of a finite reserve and saving of the surplus over and above extraction costs (rent in economic terminology), it is possible to maintain a consumption pattern indefinitely. As Common says in an eminently balanced discussion of the rule (1995, p 48), 'The Hartwick rule is to be regarded as a mathematical parable, rather than an empirical proposition about the world we live in'. He goes on to defend the model that is so important to conventional economic thinking about sustainability. He makes three broad and very useful points, some of which we have made already.

Firstly, it is anthropocentric and human consumption orientated, but ironically it tends to view the environment as purely instrumental. This raised my earlier point about fungibility in that the environment is not seen as complex and holistic.

Secondly, substitution is the key to the model but when it comes to actual substitution possibilities, which could and would unlock sustainability, it is not easy to find candidates. Economists have at times estimated elasticities of substitution between major resource and factor inputs, for example and crucially the extent to which man-made capital can replace natural capital (environmental resources) but they are not always promising or convincing. Pearce and his collaborators in his writings (for example, Pearce and Warford 1993) have emphasised this particular form of substitution, but it remains problematical. There are very few obvious examples of substitution that free up completely our reliance on natural capital and there are types of natural capital for which there are no substitutes. We will return to this issue since among traditional economic models it appears to be the most promising.

Thirdly, Common quite rightly points out that the logic of the Hartwick model is excellent. We can indeed sustain a standard of living in the face of specific resource scarcity, even depletion, if we can transform its rents into some other substitute input.

Yet there is an element of alchemy in all this that tends to produce scepticism. On the positive side however, we should note the organic nature of the market and that consumption patterns themselves will evolve both exogenously and endogenously in response to relative prices. This provides degrees of freedom to the whole process. While it smacks of hubris to assume that we can know what future generations will want and need, we can be reasonably confident that their needs will not be exactly the same as ours.

However, if we invoke Maslow's hierarchy of needs and use a little commonsense we can reasonably postulate that humans will need food, clothing, shelter and various other things into the future. Of course, as we become more affluent, the range of our wants expands enormously and seems to go into new dimensions. These are not easy to predict and become even less so if we allow that humans are a culturally and socially evolving species, as well as one that evolves biologically.

These ideas mesh well with the economists' notions of growth, increase in material wellbeing and choice, 'often almost infinite and always seen as a good thing', and thus the general idea of substitution as ever present. Indeed, economics confronted with increasing evidence of absolute scarcity, as opposed to relative scarcity, has turned to substitution as its saviour and this has led to the various concepts of sustainable *economic* development, to which we must now turn. This is entirely logical since without a resource frontier substitution is the only way out. We might note that these concepts are *economic* in that they are firmly bedded in conventional economic theory.

CONCEPTS OF SUSTAINABLE DEVELOPMENT — WEAK AND STRONG

If we return to the fundamental issue of sustainability, which is how we can live now without compromising too much the welfare of future life, (both human and we hope perhaps other life forms too), then as Turner et al. (1994) say, 'The answer is through the transfer of *capital bequests*' (the emphasis is theirs). Remember that capital for economists means resources that are capable of producing goods and services that in turn enhance wellbeing. Thus the wellbeing of future generations depends crucially on what resources we leave for these generations and in particular whether we, the present generation, exhaust some finite resources and whether we damage renewable and reproducible resources so that their future productivity is lower. The necessary and sufficient conditions for sustainable development depend on answers to these questions.

Such thinking is not new to conventional economics. Hicks, one of the fathers of modern economics and a Nobel Laureate, developed his classic concept of true economic income in his *Value and Capital* (1946). Here he argued that a proper concept of income (for consumption) is what a person or for that matter a nation, can consume within a given time period without compromising their ability to produce future income. Thus we must maintain capital and for advanced human societies, advanced both in development and thought, this means maintenance of both man-made and natural capital. Herein lies a big issue, the idea that capital can be divided usefully into these two categories. Following Pearce and Warford (1993, p 52) we can formulise this by defining the stock of all capital K as:

$$K = K_m + K_h + K_n$$

where K_m = conventional man-made capital, roads, factories, machines; k_h = the stock of knowledge, skill and technology, what has come to be called human capital and K_n = the stock of natural capital, forests, fish, oil, water, biodiversity, assimilative capacities. It is convenient for our purposes to subsume K_h into K_m since we wish to highlight the role of natural capital (the environment) in the process. Of course our main purpose is to consider substitution possibilities and these are very complex. Generally, we would expect substitution within groups to be more straightforward than between groups. Thus oil for coal rather than coal for labour or iron for water. However, this is not always so and often technology and its advances release unlikely and unexpected substitution possibilities.

Following Pearce and Warford and emphasising substitution possibilities it is useful to further subdivide the natural capital component into K_n and K_n^*, where the latter is defined as that part of natural capital for which substitutes are not clearly available or impossible. What is the purpose of this simplified formalisation of the world's capital? It is to analyse whether (this generation) any present generation, can pass on a capital stock that leaves (the next generation) future generations with production potential intact. It leads us to consider various degrees to which this condition may be met.

WEAK SUSTAINABILITY (WS)

This characterisation of the sustainability paradigm is, as its name suggests, the least demanding. It views all forms of capital as essentially similar so that it is the value of aggregate capital stock that we pass on to future generations that matters, not its composition. In effect it implies perfect substitutability between all forms of capital, which appeals to those who perhaps do not wish to second-guess future tastes and technology. Thus for example it may matter little that we use up the worlds supply of coal, if we bequeath future generations more roads. We can think of myriads of these sorts of exchanges with the future, some more bizarre than others. All that really matters are the relative values. They appeal broadly to the economists' fungible view of the world, but much less to the ecologists' and environmentalists' view that many assets, particularly natural ones, either have intrinsic value or are in some real sense special or unique.

On this basis, there has been extensive criticism of the notion of weak sustainability. Herman Daly (1995) has indeed argued that Robert Solow, Nobel Laureate, eminent economist and writer on weighty issues such as sustainability, had claimed that the world could do without natural resources. This view is gleaned from Solow's seminal piece in 1974, the date of which indicates if nothing else that good, very good, mainstream economists were thinking about these issues long before most other people were. Clearly this idea of the ultimate redundancy of natural capital invites criticism. Alan Holland, a philosopher writing in 1997, explores the extent of sustainability in detail and in essentially non-economic terms. In doing so he brings out some of the complexities and ambiguities of the issue of substitution. Without agreeing with his general conclusions, which tend to favour Solow's position that substitution has enormous potential and might indeed lead to a world greatly unreliant on natural capital, his ideas are illuminating. He explores the notion of substitution in terms of what our purpose is. For example, in eating an apple, is it taste or nutrition or both? For an economist this is reminiscent of the so-called new consumer theory of Lancaster (1969) in which he emphasises characteristics rather than goods. Thus an apple and a pear might be arranged formally in characteristics space whereas conventional economics sees them as two distinct goods.

Secondly, Holland looks at the degree of substitution. He cites the everyday idea of a 'poor substitute'. This idea is with due humility, familiar to economists and is captured in the formal idea of cross-elasticity, albeit in the context of prices. Basically, this looks at the extent to which people will substitute one commodity for another consequential upon a change in the *relative* prices of the two goods. We cannot gainsay Holland's idea, but it is scarcely new.

Holland's third question addresses the matter as to what we include as human capital. Is it all things touched by human activity, which in the modern world is arguably almost everything, or is it some much more restricted view?

All these ambiguities and definitional problems mean that actual resolution of the idea of substitution in practice are difficult. It becomes a highly empirical case-by-case issue rather than the eminently quantifiable and amenable concept that tends to pervade economic models. Once again, we get the impression that economics has had a good idea, but has perhaps simplified it beyond its usefulness. The idea that substitution is difficult to quantify in practice naturally invites us to try to refine or at least confine the idea. The concept of strong sustainability is an attempt to resolve this.

STRONG SUSTAINABILITY (SS)

Strong Sustainability emphasises the specialness of natural capital or part of it — our K_n^* in the total capital formula. What it says in essence is that some natural capital is unique in its 'function' as part of the life support system of the Earth and cannot under any circumstances be substituted so that relative valuations are irrelevant. Such things as green plants in total, water cycles and biodiversity are variously seen as *critical natural capital* (Turner et al. 1994) and as such must be retained and maintained at all costs. This type of view has achieved recognition in economic theory in the work of Wills (1997) and the concept of a safe minimum standard (SMS) and more generally in the widely pursued idea of the Precautionary Principle (see Perrings 1991). Neither of these concepts is strictly the same as the Strong Sustainability (SS) idea but they do embody notions of restraint in the use of core natural capital. They emphasise the idea of a threshold level of damage to, or reduction of, natural capital that emphasises its difference. They deny the proposition that all natural capital could be replaced by man-made capital. They lack the hubris of the Weak Sustainability (WS) ideas.

Despite the fact that SS emphasises the critical nature of *some* natural capital it does not deny the potential for substitution within the general area of natural capital. Thus it allows that natural capital takes many forms and that some are good substitutes for others. The critical issue is that we leave future generations with an unchanged quantum of natural resources K_n in total presumably including wholly intact K_n^*.

For example see Tietenberg (1994) for a definition of 'modified' sustainability — a similar concept. While these ideas are far from precise, it is clear that this idea of SS is much closer to the views of many non-economists than WS, which really is deeply bedded in conventional economic theory.

In discussing this taxonomy of sustainability concepts we must mention what is often called 'absurdly strong' sustainability (ASS), another rather unfortunate acronym. This embodies the idea that all nature and therefore natural capital is sacrosanct and can under no circumstances be substituted for. It is the preserve of deep ecologists and the like and is not vulnerable to economic reasoning. Holland (1997) again addresses these issues and points out that such a position may be defensible on two major counts. Firstly, the natural capital contains things (he calls them items), for example other higher life forms, which have moral status. This means that they cannot (and should not) be seen merely as capital.

Secondly he argues that much of the natural world 'has a special significance and importance in our lives'. Again, this sort of view is not obviously amenable to economic analysis or perhaps to any analysis, it is more an act of faith and faith is not a strong suit in economics, even though many would see neo-classical economics as a faith itself. He goes on to suggest that the natural world has its value in its 'otherness', its inherent difference from the world of man, commerce and order. We are now way beyond the limits of economics and will leave the idea of ASS!

OTHER VIEWS OF SUSTAINABILITY

Various authors (Turner et al. 1994, Jacobs 1991, Pearce 1991, Daly and Cobb 1989, Common 1995 and Solow 1991) have in their own ways attempted to outline a menu of policies that might yield a sustainable development path. None of them seem to give us an unambiguous, quantifiable and practical model with anything like the analytical and predictive power of the neo-classical model when applied to conventional markets for conventional goods. This surely is an area where we need a new economics or we must simply throw up our hands and agree that we are defeated. If sustainability is indeed a new paradigm as is argued very early in this book by Venning and Higgins (Chapter 1), then this is not surprising. In that chapter they pose economics as the traditional system and contrast this with the emerging system. Economists themselves have often recognised this dichotomy. It is reasonable to see most of Herman Daly's work as being an attempt to integrate economics and ecology, most recently through the International Society for Ecological Economics (ISEE) and its flagship journal *Ecological Economics*. Initially he attempted to do this through the idea of the steady-state economy, but this was as its name suggests a rather sterile and undynamic concept. As de Steiguer (1997) says in his essay on Daly, 'The steady-state ... insists upon a complete cessation of economic growth while sustainable development asks only for society to moderate the increase in economic growth'. Small wonder that the latter has proved to be more popular with economists and politicians.

Nevertheless, we must pause and consider Daly's contribution to the debate, which was early and profoundly important. Daly first published his ideas on the stationary state in 1971, when there was little discussion of such matters and indeed the 'first' environmental revolution of Carson (1962), Ehrlich (1968) and locally Marshall (1966) was fresh and still under way. Probably because it was economics and written by an economist, it was largely ignored. Daly himself said that even university economists had 'aggressively ignored' his ideas when writing about their impact 20 years later (1991, xii per de Steiguer 1997). Daly has always been willing to agree that the notion of a steady-state economy goes back to John Stuart Mill (1848). In Daly's hands it is best outlined in his 1977 book *Steady-State Economics* with a second edition in 1991. Overall the model, loose as it was, is of a highly constrained society much regulated and without the individual freedom of the neoclassical market. More precisely, resource throughputs were circumscribed and the main avenue for growth in the conventional sense lay with improved technology and a less well-delineated substitution of preferences towards less resource-using goods and services. To the extent that affluent societies do increase their consumption of services relative to physical goods, this substitution may be occurring, but probably not at a pace that will satisfy sustainability objectives.

In later writings and particularly in the monumental *For the Common Good* (Daly and Cobb 1989), Daly and his co-author move into more philosophical fields and stress increasingly values and *scale* and the physical volume of throughput of the economy. For example in Chapter 7 of *For the Common Good*, curiously entitled *From Chrematistics to Oikonomia*, the authors invoke Aristotle and contrast short-run economics with long-run economics or what they see as 'economics for the community', what we used to call political economy. My sympathies are with his conclusion that we need to make the market and, by close association, modern economics the servant of the community *not* its master. I believe that this is even truer 10 years or so later when market economics is if anything more revered. His view of the scale of the market is not quite the same constraint as my own view of the useful boundaries of the market (Hatch 1995) but it is similar. The two are certainly not incompatible.

Daly's contribution to the sustainability debate is perhaps more one of being a catalyst, indeed the principal catalyst among economists, rather than one of providing us with a finished model. His no net growth ideas are not and probably will never be wholly acceptable. As de Steiguer (1997, p 20) says in conclusion of his chapter on Daly, 'Therefore the steady-state economy and sustainable development are philosophies — similar philosophies in fact — more than precise models for managing the environment. Neither one specifies the exact levels of resource consumption or even the methods for calculating these levels'. De Steiguer does not bemoan this lack of a model and perhaps neither should we.

As Daly pointed out in *Ecological Economics* in 1992, economic systems should have three basic goals or features: efficient allocation, equitable distribution and sustainable scale. The first has been very adequately dealt with by economics, indeed it is its defining feature. Distribution has been a problem and though economics has flirted with it, it has never really been satisfactorily incorporated into neo-classical economics. It has always been something of an add-on. Scale, Daly's preoccupation, the physical size of environmental impact, has never really interested economists. Admittedly the eccentric Georgescu-Roegen in a long career emphasised the importance of the laws of thermodynamics and their relevance to economics (for example, 1971 and 1986). Similarly, and with more public acclaim, Boulding (1966) coined the idea of 'Spaceship Earth', which likewise drew attention to physical constraints. It is instructive to note that the idea of scale introduces above all notions of the importance of physical units and to this extent takes the emphasis away from money units and therefore from conventional economics.

Where do we stand as we enter the 21st century? Have we got a model that integrates economics with the new and fast developing 'science' of ecology and its bigger brother environmentalism? Can we move forward using the best of economic insights to help us to achieve sustainable development and perhaps growth, even in something like its conventional form?

CONCLUSION

Economics has been an enormously successful subject and has increasingly dominated politics to the extent that they are now often seen, unfortunately as coterminous. As I have suggested, the neo-classical market is often seen as the only way to organise almost everything.

Early invasions of environmental matters were seen as a natural and logical extension of the powerful tools of neo-classical economics. Beckerman and the like simply used the analysis of externalities and the various price remedies to solve environmental problems. Pollution taxes and other price adjustments seemed to deal with most environmental problems in the 1970s and even the 1980s. Thus we had the flowering of *environmental* economics in the hands of people such as Krutilla (1967), Ayres and Kneese (1969), Gordon (1954), Pearce and his various collaborators in numerous publications, for example with Markandya and Barbier (1989), and with Turner (1989), and latterly Tietenberg (1994) and in Australia, Wills (1997). This has enriched economics and much of the work has extended the subject, but little of it has actually changed the basic paradigm and the problem is that the big issues of modern environmentalism probably need this paradigm change. Many people have dubbed this new subject *ecological* economics and this has been the area of Georgescu-Roegen, Daly, Boulding and others. Again, these contributions have been enriching and very valuable but it is probably fair to say that no one has built the new economics, certainly not the tight predictive models of the old neo-classical economics. Of course the problems are infinitely more complex and by their very nature are not amenable to the simplifying assumptions that made neo-classical economics so successful. We cannot largely ignore time and ambit, and we must consider quite explicitly the rest of the living world even if only through human interests but probably in some senses, beyond them. Economics has not failed; it simply has not succeeded in doing this yet. We should not dismiss economics holus-bolus. Much of it is very useful even in trying to understand sustainable development, but it certainly cannot by itself give us the template. We, as economists, need to integrate many other disciplines, ecology, politics, law and physics among others into our models. To do this will be very difficult and may require us at times to forsake economics as we know it, but it has to be done.

Common (1995) in his rather neglected but excellent book, significantly subtitled *Limits to Economics*, faces these problems realistically. There are limits to neo-classical economics and we must recognise these. However, recognition is only the first step. The next, and much more difficult one, is to build the new economics and one of the difficulties is that this will almost certainly not be as self-contained and well delineated as the neo-classical paradigm. We must not feel that we have to defend the subject to the extent of defending its *present* boundaries. The issues are too important for this and are bigger than any discipline. As Common says (1995), 'The problem is especially difficult because the society in question is the whole of humanity. At this global level, social institutions are weak in comparison with those at the national level'. His further conclusion is that economics needs to consider criteria other than just consumer sovereignty. I certainly agree with this but wonder if economics *as we know it* can incorporate other criteria.

As stated at the outset, it is not that economics is wrong or has failed, rather that it cannot be expected to solve alone such monumental problems. There is no doubt that economics can contribute but to do so it must be willing to concede that other disciplines are relevant and indeed essential. This will be difficult for the 'social science imperialist' discipline. It will have to concede that it has no special role in deciding what the objectives should be, but as Common says, it probably can have a role in helping us to achieve the objectives. Much excellent work has been done in economics on 'instruments' or methods of achieving objectives. We can continue to use many of these, such as pollution taxes, as they have proved extremely effective in many parts of the world.

These instruments alone will however not solve the truly global problem of sustainability. This is an issue for market capitalism in the broad and therefore encompasses more than instruments. We need to consider the totality of market capitalism, which includes political structures, legal systems, education systems and more. All these things that Adam Smith implicitly knew were important as the backdrop to well-functioning markets — his 'moral sentiments'. These correctly adjusted may together with markets, provide a solution. The problem is that we cannot afford to be wrong. We certainly should not rely on markets alone. They may be one of the best ways of organising human affairs, but they are not the only way. Let us hope that this book inches us towards this complex solution to what is almost the ultimate problem for humanity.

5

ENVIRONMENTAL INDICATORS: DEVELOPMENT AND APPLICATION

JACKIE VENNING

BACKGROUND

Much of the activity in the field of environmental reporting in the late 1980s and early 1990s has reflected the growing awareness within governments and the community of the need for measures to track progress towards sustainability. The current economic reporting system fails to take account of the environment and natural resources. Environmental indicators, which simplify, quantify and communicate information, complement the current economic reporting activities. They are effective tools for monitoring and guiding change. Also they facilitate measurement of environmental performance by evaluating how well government and business are implementing environmental policies and meeting international obligations.

Through the development and application of indicators, environmental reporting is developing as a strategic tool to guide environmental management and evaluate environmental performance. The preparation of comprehensive reports, currently the mainstay of environmental reporting, should in time become a by-product of a value-adding process that is focusing on linking environmental and economic reporting in decision-making processes. The development of environmental indicators and complementary information management systems is pivotal to this evolution.

INTERNATIONAL DEVELOPMENTS

Despite the wide range of activities that are currently associated with environmental reporting, generally referred to as State of the Environment (SoE) reporting, it is still a relatively new tool for guiding resource allocation and environmental management. While some jurisdictions started reporting in the 1970s, the reports were usually thematic reports on topics such as air or water quality. The first comprehensive reports, covering all key components of the environment, did not appear until the mid-1980s and most jurisdictions only published or started preparing comprehensive reports in the 1990s. The pressure-state-response (PSR) model, now widely used as a reporting framework, was not developed until the early 1990s, and is still being refined (for example, the extension of the PSR model to the DPSIR model that is discussed in Chapter 3).

The development and adoption of the PSR model established a conceptual framework for reporting. One of the main challenges, if not the main challenge, in advancing reporting is the development of a set of agreed indicators that are widely applicable across Australian jurisdictions. However national efforts need to complement international developments, and the latter are still incomplete. Further, testing indicators for their sensitivity and reliability can take many years.

In Australia, Canada and the Netherlands, where all regions (or states) have been involved in environmental reporting, though some more recently, it is of note that there are efforts at a national level to harmonise reporting systems.

NATIONAL DEVELOPMENTS

The signing of *The National Strategy for Ecologically Sustainable Development* by Commonwealth, state and territory governments in 1992 committed all signatories to sustainable development with progress towards this goal monitored through regular SoE reports (Commonwealth of Australia 1992). This has prompted significant developments in Australia in the environmental reporting field and national SoE reporting is now mandated under the *Environment Protection and Biodiversity Conservation Act 1999*.

In Australia the PSR model, or variants of it, are used. There are benefits in having a consistent reporting system, both in terms of report structure and compatibility of data, for environmental reporting across all levels of government. The Australian and New Zealand Environment and Conservation Council (ANZECC) SoE Reporting Task Force has co-ordinated national, and state and territory, efforts to develop a set of nationally-applicable environmental indicators that will suit Australia's unique circumstances and allow comparison with the rest of the world (Table 5.1). The task force has an on-going role to review the development of indicators, in particular the methodologies and protocols for their consistent application and interpretation.

Table 5.1
Summary of national core indicators

Theme/Issue	Core Indicator	C,P,R
ATMOSPHERE		
Climate variability	Southern Oscillation Index	C
	Daily and extreme rainfall	C
	Average maximum and minimum temperatures	
Enhanced	Greenhouse gas atmospheric concentrations	C
greenhouse effect	Annual greenhouse gas emissions	P
Stratospheric ozone	Concentration of ozone depleting substances in the atmosphere	P
	Stratospheric ozone concentration	C
	Recovery and destruction of ozone depleting substances	R
	Ultra-violet radiation levels at the surface	C
Outdoor air quality	Exceedences of NEPM Air Quality Standards for carbon monoxide concentrations	C
	Exceedences of NEPM Air Quality Standards for ozone concentrations (photochemical smog)	C
	Exceedences of NEPM Air Quality Standards for lead concentrations	
	Exceedences of NEPM Air Quality Standards for nitrogen dioxide concentrations	C
	Exceedences of NEPM Air Quality Standards for sulphur dioxide concentrations	C
	Exceedences of NEPM Air Quality Standards for particles concentrations	C
	Emission of air pollutants	P
BIODIVERSITY		
Threatening	Clearing of native vegetation	P
processes	Destruction of aquatic habitat	P
	Fire regimes	P,C
	Introduced species	P
	Species outbreaks	C
Loss of biodiversity	Extinct, endangered and vulnerable species and ecological communities	C
	Extent and condition of native vegetation	C
	Extent and condition of aquatic habitats	C
	Populations of selected species	C
Biodiversity	Terrestrial protected areas	R
conservation	Marine and estuarine protected areas	R
management	Recovery plans	R
	Area revegetated	R
LAND		
Land use and management	Changes in land use	P,R
Erosion	Potential for erosion	P
	Wind erosion from high wind events	C
Salinity	Area of rising watertables	C
	Area affected by salinity	C
Acidity	Area affected by acidity	C
Contamination	Exceedences of the Maximum Residue Levels in food and produce	C

INLAND WATERS

Groundwater	Groundwater extraction versus availability	C
	Exceedences of groundwater quality guidelines	C
Surface Water	Extent of deep-rooted vegetation cover by catchment #	P
	Surface water extraction versus availability	P
	Environmental Flows Objectives	R
	Discharges from point sources	P
	Surface water salinity	C
	Exceedences of surface water quality guidelines	C
	Freshwater algal blooms	C
	Waste water treatment (inland waters)	R
	Waste water re-use (inland waters)	R
Aquatic Habitats	Vegetated streamlength	P
	River health (AUSRIVAS)	C
	Extent and condition of wetlands	C
	Estimated freshwater fish stocks	C

ESTUARIES & THE SEA

Marine habitat and biological resources	Changes in coastal use	P
	Disturbance of marine habitat	P
	Total seafood catch	P
	Estimated wild fish stocks	C
Estuarine and marine water quality	Coastal discharges	P
	Maritime pollution incidents	P
	Exceedences of marine and estuarine water quality guidelines	C
	Bio-accumulated pollutants	C
	Algal blooms in estuarine and marine environments	P,C
	Waste water treatment (coastal waters)	R
	Disturbance of potential acid sulfate soils	P
Global processes	Sea level	C
	Sea surface temperature	C

HUMAN SETTLEMENTS

Energy	Energy use*	P
	Energy sources*	P,R
Water	Exceedences of drinking water quality	C
Demographics	Urban green space	C
	Residential density	C
	Population distribution and number of people per dwelling	P
	Visitor numbers	P
Transport	Public transport use	C
	Fuel consumption per transport output	P
Waste	Solid waste generation and disposal	P
Community attitudes and actions	Community attitudes and actions	R

Key
C,P,R = condition, pressure, response type of indicator
* = also relates to enhanced greenhouse effect issue in Atmosphere theme
# = also relates to salinity issue in Land theme

SOURCE ANZECC 2000

In 1999 corporate environmental reporting became mandated under the *Company Law Review Act 1998*. Under Section 299 of this act the directors' report for a financial year must detail the organisation's performance with regard to environmental regulations in the jurisdictions within which they operate. Following the introduction of this legislation Environment Australia prepared a framework for corporate environmental reporting to assist companies with these reporting requirements (Environment Australia 2000).

One of the next steps in advancing environmental reporting as a strategic tool is the development and application of environmental performance measures (EPMs). These will enable environmental and financial reporting to be better integrated in key decision-making processes. A unified national reporting framework for Australia is important to underpin these developments.

Much work was initiated during the 1990s at a national and sectoral level (for example, agriculture, forests, fisheries) to measure sustainability. This chapter only addresses the development of environmental indicators that form a component of these measures. The development of sustainability indicators is covered in Chapter 7.

STATE AND LOCAL DEVELOPMENTS

Most states and territories in Australia have formal SoE reporting systems. In New South Wales, Queensland, South Australia, Tasmania and the Australian Capital Territory (ACT) SoE reporting is mandatory. In New South Wales SoE reporting is also mandatory at the local level under the *Local Government Act 1993* (EPA NSW 1995a). Details of specific state programs are included in Boxes 5.1 and 5.2.

All states and territories with regular reporting systems structure their reports around the key components of the environment (or themes), for example, atmosphere, water, land and biodiversity, as in the 1996 Australian SoE report (see SEAC 1996). Theme headings can be further subdivided by environmental issues (EPA 1998) or by region (Government of Western Australia 1998). Some reports also include sections that address the impacts and management responses specifically relating to key sectors, for example, mining, agriculture, forestry, energy, in separate sections (SDAC 1997, Government of Western Australia 1998).

As information is often lacking on the contribution of sectors, such as fisheries, forestry, agriculture, transport and energy, to environmental pressures, the preferred trend is to integrate available data into chapters covering key components of the environment. For example, the New South Wales 1997 SoE report contained only five chapters: Air, Land, Water, Biodiversity, Towards Sustainability and an introductory chapter, rather than the 23 chapters included in its 1995 report (EPA NSW 1995b, 1997b).

In the ACT, Tasmania and Western Australia the SoE reporting process distributes reports calling for comment from the Government and the community. In the ACT responses are tabled in Parliament. In Western Australia the Government produced a separate report detailing its actions in response to the SoE report (Government of Western Australia 1999).

DEVELOPMENT OF INDICATORS

Evaluation of early SoE reports demonstrated the difficulty of evaluating the effectiveness of many environmental and natural resource management programs. While reports continued to be produced without agreed key indicators and performance measures any environmental reporting process would remain qualitative, rather than quantitative, and therefore largely subjective. By comparison one of the strengths of economic reporting is the development and adoption of indicators that are applicable across different jurisdictions and at varying scales.

Before developing indicators it is important to determine the purpose for reporting, a conceptual framework and selection criteria (EPA NSW 1996, EPA 1997). The purpose of SoE reporting is variously defined, but in the main, reports are intended to provide credible and quantifiable information about the quality of the environment and quality and quantity of natural resources. The purpose of sustainability or quality of life reports, which usually contain a suite of economic, social and environmental measures, is to assess whether we are maintaining our economic prosperity and quality of life while at the same time maintaining the ecological processes on which life-support systems depend.

CONCEPTUAL FRAMEWORK

The Organisation for Economic Co-operation and Development (OECD) pressure-state-response (PSR) model has been widely applied and variously adopted in SoE reporting. It is based on the concept of causality (refer Figure 5.1). Humans exert pressures on the environment through their activities and change the quantity and quality of natural resources (state). Society responds to these changes through implementation of environmental policies, economic investment, and research and development to ameliorate pressures on the environment. The model takes an entirely anthropocentric view of pressures on the environment.

In countries such as Australia, where climate variation is comparatively high, the model requires that natural variability be incorporated into the state component rather than being treated as a pressure. Despite limitations of the PSR model it has become widely adopted as the basis for SoE reporting and, as in any reporting system, consistency in approach is essential across jurisdictions. In the 1996 national SoE report, the pressure-state-response framework was used. A review of the 1996 report proposed that future reports use a modified condition-pressure-response framework to focus attention on the condition

Box 5.1
SoE Reporting in South Australia • Case study 1

The first comprehensive South Australian SoE report *The State of the Environment Report for South Australia* was published in 1988 (EPC 1988). At that time there was an understanding that reports would be produced every five years. The 1988 report followed the format of then OECD reports and covered both key environmental resources and sectors (in all 15 chapters). Comprehensive SoE reports have since been produced in 1993 and 1998.

A state-based appraisal of the SoE reporting process in 1994 demonstrated the difficulty of evaluating change and the effectiveness of many environmental management programs without agreed key indicators and performance measures. There was concern that while we continued to report without such measures any environmental reporting process would remain discursive and therefore largely subjective. There was then, and still is, a need to collect and manage key sets of data as a basis for regular reporting. In defining those measures around which to frame a future reporting system, the challenge is not to compile more datasets but to identify the key datasets to monitor the state of our natural resources and ecosystem processes.

Since the earlier reports were prepared more stringent reporting requirements have been legislated in keeping with the *National Strategy for Ecologically Sustainable Development*. In South Australia requirements for SoE reporting are contained in Section 112 of the *Environment Protection Act 1993*. This states that reports must be produced by the Environment Protection Authority at least every five years and:

- include an assessment of the condition of the major environmental resources of the State
- identify significant trends in environmental quality based on an analysis of indicators
- review significant programs, activities and achievements of public authorities relating to the protection, restoration or enhancement of the environment
- review the progress made towards achieving the objects of the Act
- identify any significant issues and make any recommendations requiring the attention of the Minister.

The *State of the Environment Report for South Australia 1998* (EPA 1998), the third in the series of comprehensive reports, was compiled around a set of predefined environmental indicators. Following its release the indicators were reviewed and updated and 76 key environmental indicators were selected to report on changes in the quality of the environment, and quantity and quality of the State's natural resources. A set of 'headline' indicators was selected and for these a set of environmental reporting measures (EPMs) were proposed (Appendix 5.1). For these, the ▶

of the environment rather than pressures on the environment. In the 2001 report the model will be expanded to include an analysis of implications and the report will be based on a condition-pressure-implications-response framework. This fits somewhere between the PSR and DPSIR models discussed in Chapter 3.

▸ data must be more regularly collected to enable timely reporting and establish the necessary links with key decision-making tools and processes, for example, financial and economic reporting.

The EPMs were set in accordance with the principals of ecologically sustainable development and aim to halt further decline of the State's natural resources. (DEHAA 1999). On the whole public response to the EPMs was positive. However one criticism was that they set the benchmark too low. Rather than slow or halt further degradation of our resource base it was argued they should be set to reverse and ameliorate degradation.

It is anticipated that this set of 'headline' indicators will form the basis for more regular reporting. It is envisaged that the adoption of these performance measures will see South Australia well placed to:

- withstand external scrutiny of the 'clean and green' image being promoted to interstate and overseas markets;
- demonstrate to overseas markets that the use of our renewable resources is sustainable;
- demonstrate that we are working to achieve the ecological outcomes that the community demands;
- make well-informed decisions concerning the allocation, use and management of natural resources.

A departmental task force was established in October 1999 to advance the integration of performance measures into its corporate business. However in the absence of whole-of-government outcome statements the application of EPMs will be limited.

Figure 5.1

The pressure-state-response model
SOURCE SoE Unit, Environment Australia (adapted from OECD)

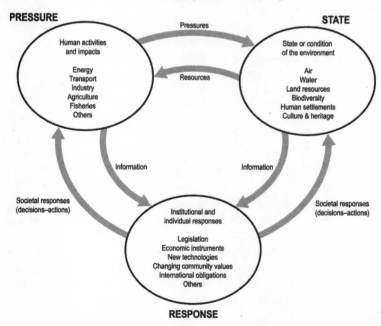

Box 5.2
SoE Reporting in Western Australia • Case study 2

The first comprehensive Western Australian SoE report was produced by government agencies in 1992. At that time there was an expectation that a further report would follow in four years' time. Consequently in 1995 the Western Australian Government initiated the process for a second report. The Minister for Environment, with Cabinet approval, established a SoE Reference Group comprising community and government representatives with terms of reference to:

- co-ordinate the production of the report;

- identify key environmental issues by region prioritised for government and community action;

- develop a framework for on-going reporting including objectives, benchmarks, indicators and data assembly.

Regional Focus Groups were established (eight marine and 15 terrestrial) based on environmental regions across the State. These focus groups compiled inputs from the general public, individuals, industry and government. The information was discussed by experts and formed into eight SoE Reference Group Draft Working Papers. First generation indicators were included in the working papers.

In 1997 the SoE Reference Group condensed the information in the working papers into a concise draft SoE report that was submitted for review by eminent scientists. The Minister for Environment then released *Environment Western Australia. 1997 Draft State of Environment Report* (Government of Western Australia 1997) for public comment. The draft report was very much aimed at decision makers within government and the broader community. Its format was designed to be succinct and easy-to-read, and it extended the condition-pressure-response approach by including sections on current responses and implications. ▶

NATURE OF INDICATORS

An indicator is a physical, chemical, biological, social or economic variable that points to significant outcomes and can be used for management purposes. Indicators differ from other measures in providing meaning beyond the attributes directly associated with them, either by comparison to a standard or a reference value (DEST 1994). An indicator within the context of SoE reporting, is a measure that describes the state of the environment or provides information about a phenomenon. The OECD framework identifies three categories of environmental indicators (OECD 1994):

- indicators of environmental pressures describe pressures from human activities on the environment;

- indicators of environmental condition relate to the quality of the environment and the quantity and quality of natural resources;

- indicators of societal responses refer to individual and collective actions to reverse, ameliorate or prevent human-induced impacts on the environment and to conserve nature and natural resources.

▸ A 'road show' was conducted to raise the profile of the draft report in the regions and to encourage public comment following its release. There was significant public response with over 500 points of contention and the draft report was revised on the basis of these comments. A document was also published summarising public comment and how it was addressed in the revised report.

The Minister for Environment released *Environment Western Australia 1998 State of Environment Report* (Government of Western Australia 1998). This final report included environmental objectives and indicators and further enhanced the response section by including suggested responses. The report was aimed at directing policy changes in the environment and made more than 100 suggestions for new initiatives. On release of the report the Government made a commitment to respond to it. A Cabinet approved process was put in place across government bringing chief executive officers together from over 20 agencies to address jointly the issues raised and to reply to the suggested responses in the report.

In 1999 *Environmental Action: Government's Response to the State of Environment Report* (Government of Western Australia 1999) was released. This report is an environmental action plan by State Government and details some 180 actions. The challenge is for other sections of the Western Australian community to respond as well.

Subsequently an evaluation of the SoE reporting system was conducted using independent assessors. The evaluation had two purposes. Firstly, to inform the design of the next state of environment reporting cycle, and secondly, to make open and transparent the decision-making processes within the environment reporting program. The next round of reporting, to begin in 2001, will incorporate aspects of the evaluation findings and will examine the effectiveness of actions taken by Government and the community.

* Text provided by Ray Wallis, SoE Reporting Unit, Western Australian Department of Environmental Protection

The reporting framework (or reporting matrix) is completed by the identification of environmental themes or issues, either by source or sector (see Appendix 5.1). These are chosen to reflect current environmental concerns and therefore can vary over time reflecting the inclusion of new issues or exclusion of old ones. An issue-based approach to indicator selection can focus reports only on matters of current concern and result in emerging issues being overlooked. In the process of indicator selection all the key natural resources (or sources) should be covered as a matter of course. This should facilitate evaluation of changes in the quality and quantity of the environment and natural resources over time and thus an assessment of how well they are being managed. The purpose of the report is important here in guiding indicator selection.

The OECD reporting format has been widely applied. It has adopted a source-sector approach to complement the PSR model and complete the matrix of indicators. However addressing the impacts of key sectors in separate chapters can fragment information and result in repetition. For this reason there has been a trend toward integrating sectoral information into the chapters on key components of the environment (otherwise referred to as themes or sources). In the national reporting framework, issues are considered and the suitability of data evaluated within the context of the following themes: Atmosphere, Inland Waters, Coasts and the Sea, Land, Biodiversity, Human Settlements, and Natural and Cultural Heritage.

The concept of macroindicators has been applied in the development of indicators to measure progress towards sustainability (Australian Bureau of Statistics 2000). They have been suggested as a more practical means of gauging progress towards sustainability than the development of composite sets of measures (or indices). They are single measures, which by their very nature integrate a lot of information about the condition of the environment. For example, salinity of the Murray as measured at Morgan, South Australia, not only tells us much about the management of the river system, but also provides an important message about the likely impact of trends for Adelaide metropolitan water supplies in the early decades of the 21st century (pers. com., Allan Haines, Environment Australia). Other measures include greenhouse gas emissions per gross domestic product and extent and condition of remnant native vegetation.

At the current stage in the development of environmental indicators most of the indicators are simple measures relating to a single aspect of the environment. Even indices that have been developed, for example, headline sustainability indicators (Australian Bureau of Statistics 2000), sustainable agriculture indicators (ARMCANZ 1998), are compilations of simple measures covering a range of factors important in the overall analysis of sustainability. The desire to develop a single measure (or composite measure) to illustrate the condition of our environment, or to demonstrate that we are managing our resources sustainably, is very appealing, particularly for capturing media headlines or for simplifying large amounts of complex information in order to convey a general impression of progress. A composite measure is developed by the aggregation of two or more variables or indicators. However, equally, there is strong opposition to the use of a single measure for monitoring environmental trends or progress towards sustainability. It is argued that single measures can be misleading because of the subjective weighting of various criteria, the aggregation of like and unlike, and the masking of detail.

In the early 1990s Adriaanse developed key environmental indicators for monitoring progress towards sustainability in the Netherlands, some of which are composite measures (Adriaanse 1993). In monitoring the Dutch policy to reduce greenhouse gases by more than 50 per cent from 1988 levels by 2020 he used a weighted summation of the Dutch annual discharge of carbon dioxide, methane and nitrous oxide expressed as CO_2 equivalents.

In February 1996 the New South Wales Environment Protection Authority held a workshop in Sydney to advance the development of composite measures for SoE reporting purposes in Australia (Harding and Eckstein 1996). Also, in February of that year a workshop was held in Adelaide to discuss indicator developments for SoE reporting. A number of the indicators identified as desirable but in need of further work were composite measures relating to the health of environmental systems and included (Venning 1996):

- Index of estuarine health. The identification of this index reflected the importance of developing broadly-based measures of ecosystem health.

- Index of vegetation condition (or ecosystem condition). The area of remnant vegetation and its condition is a potentially useful measure of biodiversity. While techniques had been developed for assessing the condition of rangelands vegetation at that time no similar work had been undertaken on a state-wide basis for assessing the condition of remnant vegetation in the agricultural regions.

However the ANZECC SoE Reporting Task Force considered it was premature to progress work with composite measures in the absence of a nationally-agreed set of environmental measures. As a consequence little, if any, work has been progressed by the states and territories towards the development of composite indicators. More recently some developmental work towards composite measures for estuarine, catchment and landscape health has been progress under the National Land and Water Resources Audit.

CRITERIA FOR SELECTION

The OECD has identified a number of criteria for indicators (OECD 1994). They should be:

- able to show trends over time;

- simple, well founded in technical and scientific terms and based on internationally-accepted standards;

- easy to interpret, readily available, adequately documented, of known quality and regularly updated using reliable procedures;

- national in scope or at least able to relate to nationally significant regional issues;

- availability of data at a time suitable for reporting purposes;

- cost effective to update regularly;

- suitable for evaluating environmental performance.

ANZECC (2000) identified a number of criteria for indicator selection and these closely mirror those developed and applied by the OECD in that they reflect the need for relevance, reliability and timeliness. The core set of national indicators were selected on the basis that they:

- reflect a valued element of the environment or an important environmental issue;

- have relevance to policy and management needs;

- be useful for tracking environmental trends at a range of spatial scales from the local to the continental;

- be scientifically credible;

- be cost effective;

- serve as a robust indicator of environmental change;

- be readily interpretable;

- be monitored regularly, either by existing programs or by new programs that might be established in the future at reasonable cost;

- reflect national programs and policies.

Ideally, measures selected should satisfy all criteria. However, often indicators are selected despite not fully satisfying all criteria because they rate highly against some 'weighted' criteria. For example, the quality of data and its accessibility is often a pragmatic determinant of selection. There is little point is developing indicators if there is little, or no, data to support their use for regular reporting. Data needs to be 'readily available, adequately documented, of known quality and regularly updated using reliable procedures'.

In effect, these criteria favour parameters for which good time series data are available. However, even the better data for reporting purposes are often of variable quality. Consequently it may be necessary to include parameters for which lesser quality data, and in some instances no time series data are available, in order to cover all key elements of the environment. For example, a measure of biodiversity (the variety of all life forms including ecosystem, species and genetic diversity) is not yet possible with current information. However various attempts have been made to select measures that will provide an indication of current trends in biodiversity. These measures are called surrogates. The condition indicator 'populations of selected species' was identified as a surrogate measure in the ANZECC national core set as changes in abundance and distribution may be indicative of general trends at an ecosystem or genetic level. The extent of remnant vegetation and its condition has also been suggested as a surrogate measure of biodiversity. In many instances though these surrogate measures also lack sufficient data to generate trends or are limited in their spatial coverage.

When selecting the national set of measures not all of the environmental indicators currently in use were deemed suitable as core indicators. For example, some measures are important for a particular purpose in a specific region but may lack national significance.

SCALE AND INTERPRETATION

Different scales are appropriate for the measurement and interpretation of different indicators. For example, data for air quality is collected by 'airshed', for biodiversity by biogeographic region (for example, IBRA), for marine and coastal areas by biogeographic region (for example, IMCRA), for land use and management by local catchment areas or land use categories depending on the purpose of the report and for population trends by the ABS statistical divisions or in more specific instances by postcode. Therefore the purpose of the report, for example, Local Agenda 21 or national SoE reporting, determines the nature of the indicators selected.

Some indicators are generic by their very nature and lend themselves to interpretation across a wide range of scales, for example, local, regional, state and national. Indicators such as 'annual greenhouse gas emissions' and 'extent of seagrass meadows' can be aggregated and disaggregated for reporting at various scales. However indicators such as 'extinct, endangered and vulnerable species', despite being one of the most widely used biodiversity measures, are difficult to interpret and cannot be readily aggregated or disaggregated. A species given a 'threatened' rating in a region may be rated as common at a state or national scale. The use of threatened species or ecosystems as a measure of biodiversity also raises the issue of interpretation of information.

Many Australian native species are naturally rare due to their restricted distribution and a 'threatened' rating may not necessarily be indicative of any change in status. Further there are frequently changes in the threatened listings (particularly for bird species) that often appear to reflect changes in bird observations and researchers' interpretations of these data rather than a change in the status of species *per se*. Despite the difficulties with interpreting this information any report on biodiversity would be considered incomplete without this information, despite its limitations. Ideally indicators should be chosen that do not have such fundamental difficulties in interpretation.

Where there are difficulties in interpretation the text accompanying an indicator should set out the limitations of the data, the significance of the trend observed and its meaning. Interpretation within the Australian context can be made difficult due to the large climatic and geographic variability and may require separating the long-term trend from the noise (short-term variability). Averaging out may not be an option in presenting the data particularly where the extent of the variability is an important consideration.

DATA QUALITY AND ACCESSIBILITY

While there are no absolute measures for data quality, widely-accepted attributes include relevance, reliability and timeliness; criteria linked to indicator selection as previously outlined. It has been argued that the key criterion in indicator selection is data availability. Data considered for selection comes from one of the follow categories (OECD 1994):

- data readily available;

- data available but needing additional work (verification of data, patchy or limited coverage);

- little or no data available and therefore not suitable for reporting purposes in the short to medium term.

Invariably measures included are those from the first category. It has been argued that rather than report on what we have, we should be endeavouring to identify what we need to report on, and then developing the monitoring programs to support reporting requirements. The ANZECC process endeavoured to address this issue by selecting indicators that were considered to be important for addressing key environmental issues but were still in need of further work (identified in the report as indicators for second stage implementation). These included:

- Extent and condition of native vegetation
  Information exists for measuring the extent of terrestrial native vegetation communities but relatively few data are available for aquatic vegetation. Methodology for assessing condition requires further development and therefore was noted as an indicator for second stage implementation.

- Populations of selected species
  Considerable research is needed to identify species that are effective indicators of change in biodiversity. Until methodology is developed this indicator will have limited application.

- Protected areas (both terrestrial and coastal)
  Methodology needs development for those areas that are outside the IUCN classifications system for protected areas.

- Coastal discharges
  National Pollution Inventory (NPI) methodology can be used to provide data for point sources and is the first stage of implementation for this measure. NPI methodology for non-point sources is under development and will represent the second stage of implementation for this indicator.

However where data was lacking or methodology needed development the indicators were still included with appropriate caveats. Data collection programs are often resource intensive and unless the data is required under legislation invariably fall off the end of the list of priorities. Efforts have been made to establish community-monitoring programs to collect data for research and monitoring purposes, for example, Frog Watch and Salt Watch, with mixed success. While involving the community has a number of benefits, the challenge is to maintain support for these programs over the long term to obtain meaningful time-series data.

The issue in environment reporting is not necessarily the quantity of data but its quality and availability for reporting on changes in the condition of the environment. This has limited the timeliness and responsiveness of information delivery. The requirement for delivery of information through media, such as interactive information systems now available through the World Wide Web, will necessitate improved data recording and management protocols. The development of interactive environmental information systems, such as those that have been developed for the national pollution inventory held by United States Environment Protection Authority, will advance access and analysis of those datasets used as a basis for reporting.

KEY ENVIRONMENTAL INDICATORS

The key (or core) measures should permit an assessment of the condition of the major environmental resources and identify significant trends in environmental quality. They should be those that most appropriately reflect changes in the issue under consideration but also be the minimal set needed so as to provide a long-term focus and avoid information overload. Most key indicators are pressure or state indicators (refer Table 5.1 and Appendix 5.1). Response indicators are needed though to complete the PSR framework and thereby permit evaluation of the effectiveness of response mechanisms to observed environmental trends.

The relatively few response measures reflect the challenge in trying to develop measures that can be directly related in a meaningful way to changes in either condition or pressure. Many measures to monitor performance have a long lag time or lack a one-to-one relationship between action and response. Response measures often used relate to 'numbers of activities', for example, number of management plans, amount of funds, and these do not necessarily reflect a direct relationship with changes in the environment or the resource being managed. As an example, the existence of a management plan does not necessarily reflect amelioration of pressures or an improvement in condition. Similarly for dollars spent. The situation is further compounded by the fact that funds well spent may not necessarily correlate with ameliora-tion of pressures and improvement of condition within the timeframe for reporting due to a lag time (often spanning decades) between action and response.

REPORTING CYCLES

At present environment reporting cycles vary from three to five years and are not synchronised. A reporting system more in line with finan-cial reporting would help to make environment information more accessible and more relevant. This would necessitate more concise reports being prepared around an agreed set of core indicators on a much shorter reporting cycle than every three to five years as is now commonly the case.

More regular reporting is contingent on the establishment of improved information systems. However it is not only a question of improvement data quality and accessibility. The nature of the data itself needs to be considered and our ability to interpret changes in the data in a meaningful way over shorter reporting periods. The SoE indicators currently used fall broadly into two timeframes for meaningful inter-pretation and evaluation:

- 1–2 year timeframe for resource consumption and pollution data, for exam-ple, exceedences of NEPM guidelines, groundwater allocation and extraction, drinking water quality, per capita water consumption, energy use, energy intensity, solid waste to landfill, amount of waste recycled, litter surveys.

- 3–5+ year timeframe for measures reflecting ecosystem health, for example, changes in sea level, changes in atmospheric concentration of greenhouse gases, area and condition of wetlands, extent and condition of remnant terrestrial and marine vegetation, numbers of threatened species, population trends for key species.

One component of quality reporting is timeliness of information for the purposes required. Shorter reporting cycles would bring environ-mental reporting more into line with economic reporting, facilitating the inclusion of environmental information in the decision-making process and enabling consideration of economic, social and environ-mental aspects. At present one of the factors limiting regular environ-mental reporting is not so much the amount of data available but its accessibility. There is a need to establish information systems within organisations to facilitate ready access to and interpretation of key environmental data sets. Better access to information would assist inte-gration of environmental reporting into key public and private sector decision-making processes.

Preparation of the South Australian Business Vision 2010 annual 'state of the state' reports (SABV 1999, 2000) illustrate some of the limitations of shorter reporting cycles for reporting on the state of the environment. These reports contain a mix of economic, social and environmental indicators. When compiling the 2000 report the environmental indicators had to be reviewed and some dropped. There were two reasons for this. Data custodians were not able to supply verified data within the short time lines available for report preparation and for some measures the change in trends over such a relatively short period were not meaningful.

Table 5.2
Environmental performance measures

SOURCE Australian National Audit Office (1996), DEHAA (1999)

Type of EPM	Description
Baseline	Often measurement of change from a pristine or near pristine environment or change since a stated time. For example, a comparison of existing vegetation types with maps of vegetation types pre-1750 to determine extent of clearing and modification since European settlement (Saunders et al. 1998).
Benchmark	A value that has some predefined environmental significance (scientific) or that demonstrates achievement of best practice (corporate). For example, NEPC guidelines for ambient concentrations of total suspended particulates, total suspended particulate lead, PM10 — inhalable, carbon monoxide, ground level ozone, nitrogen dioxide (NEPC 1998).
Standard	A predefined level of excellence or performance specifications that can be set for inputs, processes, outputs or outcomes. For example, the International Standards Organisations (ISO) has developed a series of standards (ISO 14001, 14004, 14024, 14031, 14040) to help organisations manage the impact of their activities on the environment (EPA NSW 1997a, EA 2000).
Target	Quantifiable levels or ranges to be met at a specified future date. For example, Kyoto Protocol that sets CO_2 emissions for Australia at 108 per cent of 1990 baseline level by 2008 reporting period for example, 50 per cent reduction of solid waste to landfill per capita from 1990 base by 2000 (EPA 1994).

MEASURING ENVIRONMENTAL PERFORMANCE

While environmental indicators are useful for depicting trends in the quantity and quality of a jurisdiction's natural resources, alone they do not necessarily provide the level of information to guide policy and evaluate programs. The adoption of targets and other performance measures are needed to complement them. The Montreal Protocol demonstrates this point. The phasing out of ozone-depleting substances in line with targets agreed under this protocol has seen a levelling off of atmospheric concentrations of CFCs.

Performance measurement can take place at a number of levels within and between organisations and can be made against baseline data, benchmarks or targets (Table 5.2). Not only do these measures allow more effective monitoring of progress in environmental management, they also enable environmental concerns to be integrated into sectoral policies and add economic dimensions to environmental policy performance.

Benchmarks and targets have already been set for a number of international (for example, Montreal Protocol, Kyoto Protocol) and national (for example, ANZECC water quality guidelines) issues. However there is still considerable work to be completed at state and local levels to establish performance measures that will enable government programs to be framed within limits of sustainability. In the absence of agreed targets, measures should be set in accordance with the principles of sustainability (the principles of sustainability are discussed in Chapter 2).

In South Australia a set of 25 indicators has been developed as a basis for more regular reporting on environmental performance (DEHAA 1999). These measures were nominated in accordance with the principles of ecological sustainability to demonstrate in time that the use of the State's renewable resources is sustainable. In several instances it was not feasible to set targets within strict numerical limits. For example, the target for groundwater simply states that the groundwater allocated should be less than or equal to recharge (see Appendix 5.1).

These 'headline' measures have been selected to facilitate more frequent reporting and enable environmental information to be more readily integrated into decision-making processes. However this is contingent on the establishment of information systems by data custodians that will facilitate access to key environmental data sets. Measuring progress against or towards a desired end point requires good time series data in order to evaluate progress over time.

AT THE MICRO-LEVEL

Environmental concerns, growing public pressure and regulatory measures are changing the way people do business around the world. Moves towards triple bottom line reporting (financial, social and environmental) by large companies in Australia are evidence of this. Consumers and shareholders are increasingly demanding environmentally-friendly products and services that are delivered by socially responsible companies. It is becoming increasingly important for organisations to demonstrate that not only their philosophies but also their investment strategies and day-to-day operations are sustainable. The number of annual environmental reports being published by large national and multi-national corporations indicates that sustainable development is becoming an integral part of their corporate operations (for example, Unilever 1998, The Body Shop 1999, Western Mining Corporation Limited 2000). Not only are corporations now setting environmental goals but also some have their performance with regard to these goals audited by an independent body. While at present this is the exception rather than the norm it is the beginning of a trend that is anticipated to grow.

Many companies are establishing environmental management systems (EMSs) that conform to ISO 14000 guidelines in order to remain competitive in the global marketplace. For many companies, their competitors are seeking ISO 14001 registration and their customers are beginning to look for compliance with these guidelines. An EMS that complies with ISO 14001 will ensure that a company has an effective environmental management program that can:

- reduce consumption of materials and energy in the production of goods and services;

- reduce cost of waste management;

- lower distribution costs;

- improve corporate image and customer loyalty;

- provide a framework for continuous improvement of environmental performance.

The standard ISO 14031 gives guidance on the design and use of environmental performance evaluation within an organisation. It proposes two types of indicators: environmental condition indicators, for example, air and water quality, and environmental performance indicators, for example, use of energy and natural resources, emissions and wastes. Despite calls for ISO 14031 to establish a standard set of indicators it leaves organisations free to decide which indicators to use and how to measure them.

The benefits of a generic set of environmental indicators that are widely understood and generally applied are becoming increasingly evident. State and local governments have a key role to play in establishing overarching reporting systems. In the absence of industry standards for environmental performance levels, overarching government reporting systems can be a useful guide to companies and industries seeking to monitor and demonstrate their own environmental performance.

AT THE MACRO-LEVEL

The development and application of EPMs establish high order outcomes for the environment portfolio against which program outcomes within and across public and private sector organisations can be compared. This improves transparency and accountability in decision making. The development of management frameworks to evaluate performance enables the price and quality of services to be benchmarked against best practice. Figure 5.2 illustrates a framework to integrate project performance (for example, outputs, outcomes) with broader reporting functions (for example, SoE reporting) through the development and adoption of performance measures.

Figure 5.2
Macro and micro-scale performance reporting

SOURCE Adapted from Australian National Audit Office (1996) and South Australian Department of Treasury and Finance (1998)

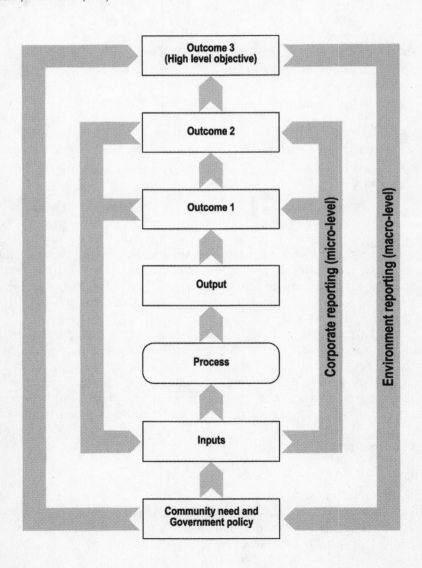

INTEGRATING AT THE MICRO- AND MACRO- LEVELS

The adoption of performance measures establishes a direct link between environmental reporting and strategic planning processes within public and private organisations (Figure 5.3). Outputs from projects or programs can be evaluated against desired environmental outcomes. Regular comparison of achievements against environmental performance measures provides opportunity for review and refinement to achieve best practice. Project activities can be constantly refined to ensure they are working towards desired environmental outcomes.

Figure 5.3
Integrating environmental reporting into the strategic planning and budget planning cycles

SOURCE Adapted from Australian Local Government Association (1999)

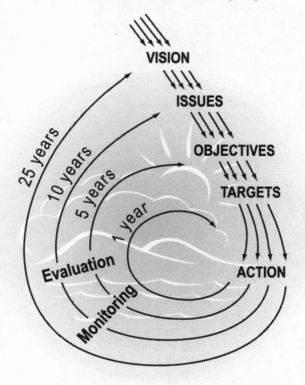

Using performance measures at various levels (for example, outputs, outcomes) allows comparison of individual and collective contribution towards environmental goals and thereby the effectiveness of environmental policy. As environmental reporting systems standardise on what gets measured and how, and as environmental performance measures are more widely developed, organisations and the community will be able to compare the performance of individual organisations against overall environmental trends. Within industry this will move the focus from compliance to resource efficiency, pollution prevention and product stewardship.

In support of this evolving reporting system, increasingly sophisticated information technology will be able to rationalise and integrate disparate datasets within a single framework. Standardisation of environmental information facilities at national and international levels (for example, National Pollution Inventories) will facilitate international comparisons and benchmarking of performance.

INTEGRATING ENVIRONMENTAL AND FINANCIAL REPORTING

The establishment of environmental performance measures provides links with the financial reporting cycle. It enables environmental outcomes to be reviewed against expenditure and provides a mechanism to allocate appropriate levels of resourcing for environmental management programs. Increasingly sophisticated information technology will serve to rationalise and integrate disparate data sets within a single reporting framework.

However in the environment and natural resource portfolios there can be a lag time between the taking of actions and the achievement of desired outcomes that exceeds the short to medium term planning associated with financial reporting. So while it can be argued that it is possible to evaluate environmental outcomes against financial reporting cycles it needs to be done in the knowledge that many environmental trends are long term rather than short or medium term. It is possible to take effective and efficient action to ameliorate environmental issues but not observe the desired outcomes for some years or even decades. This indicates the need for expert analysis of the data when interpreting and reviewing environmental performance information.

BENCHMARKING PERFORMANCE

Benchmarking performance provides guidance for improvement to achieve best practice. Even though there is worldwide interest in benchmarking environmental performance it is not possible yet for a number of reasons. These include the lack of progress in developing environmental performance indicators and the absence of any generally accepted standards across the private and public sectors for environmental reporting. As environmental performance information expands and becomes more consistent in quality and scope, governments, industry associations and non-government organisations will be able to compare the performance of individual organisations against overall trends. Performance measures enable a level of transparency in environmental reporting not previously possible.

Since the late 1980s environmental reporting has developed rapidly in response to the need to track progress towards sustainable development. Public and private sector organisations are now endeavouring to demonstrate that their activities are sustainable by monitoring their environmental performance. Standardisation of environmental information at national and international levels will eventually facilitate comparisons and ultimately the benchmarking of environmental performance.

FUTURE DIRECTIONS

LINKS BETWEEN GOVERNMENT AND CORPORATE REPORTING

At present the government reporting processes, referred to as SoE reporting, are run separately from corporate reporting (or public environment reporting) processes even where the same government agency has responsibility for both functions. There could be benefits in aligning these two processes using core environmental indicators that have widely-accepted meaning and widespread application. It is of note that in the 'State of the Economy' reporting process there is far greater applicability of measures that are more specifically applied and better linked to policy making. Such should ultimately apply for environment reporting if it is to 'come of age' and perform a fundamental function in decision making.

ENVIRONMENT 'BULLETINS'

The ultimate success of environmental reporting will depend on its ability to meet the needs of key stakeholders. In the Information Era decision makers expect timely and responsive reporting mechanisms that provide easy access to the information they need to meet community expectations of transparency and accountability.

Environmental information relating to land and biodiversity spans longer timeframes than is considered ideal for many key decision-making processes that are more closely aligned to economic and financial reporting timeframes. Clearly this is a challenge to be met if environmental reporting is going to become pivotal in mainstream decision-making processes. The use of interactive reporting online is one mechanism to address the issue of timeliness and responsiveness but it requires a level of data quality and accessibility that is not currently the norm in many government and corporate organisations.

An online report from British Columbia gives some indication of the likely nature of environment reporting in the future (Ministry of Environment, Lands and Parks 1998). It is a tiered reporting system and provides considerable flexibility in presenting up-to-date and relevant information. In online reports information can be layered with up to four (possibly more) layers being accessible.

- Layer 1
  This comprises a list of the key issues and some 'dot points' illustrating what is addressed for each issue.

- Layer 2
  This contains a brief description of why the issue is important and what has been achieved in this area since the last reporting period. The text is accompanied by a graph, often of time series data, illustrating key trends. Links can be provided to other websites providing further information about the issue. An icon located under the graph, or graphic, can direct the reader to the original data.

- Layer 3
  Tables can be included containing the information on which the graph has been drafted. This layer can also contain original data, location maps or metadata.

- Layer 4
  This level can provide the reader with direct access to the original data collections held by government or research agencies. Access to this layer requires appropriate security provisions.

This type of layered interactive reporting in effect enables users to create their own environment report. Further the most up-to-date information can be incorporated as soon as it becomes available. Hard-copy reports produced once every 3–5 years, as is currently common practice in Australia, do not provide anywhere near the same degree of flexibility.

DECISION-SUPPORT SYSTEMS

This is in essence a system much more closely aligned with economic reporting processes and underpinning sustainability reporting.

The development of sound monitoring and reporting practices is vital for effective environmental management. A key step towards effective reporting has been the development of environmental indicators to act as a baseline against which to monitor future environmental trends. Significant mistakes and environmental decline might have been avoided if such gauges had been used to drive past policy development and decision making. The challenge is to apply these measures to avoid further environmental decline and bring about improvements in our institutionalised decision-making processes.

The establishment of environmental indicators for reporting has illustrated the paucity, and all too often the inadequacy, of the data to underpin well-established reporting processes. The selection of measures as a basis for informing sustainability should not be constrained by what we are currently measuring. A long-term commitment is needed to maintain data collecting processes for core data sets at both state and federal levels. The National Land and Water Resource Audit offers the opportunity to drive much needed data collection Australia-wide.

Ultimately the data sets need to be readily accessible and linked through inventories and search facilities such as the Australian Spatial Data Directory. This will enable more timely and up-to-date reporting through online facilities. Increased access to current data management facilities could foster greater public sector involvement in an area that has previously been the precinct of government and research organisations. The ultimate challenge will be to integrate disparate economic, social and environmental data sets into decision support systems to place a higher value on our environment and natural resources and achieve sustainable outcomes.

Appendix 5.1
Core environmental indicators and performance measures proposed for the South Australian SoE reporting framework

SOURCE DEHAA 1999

ISSUE	INDICATORS	TYPE
THEME: ATMOSPHERE		
Metropolitan air quality	Number of exceedences of NHMRC or NEPM targets	C
Ozone depleting substances	Atmospheric concentrations of ozone depleting substances	P
Enhanced greenhouse effect	Greenhouse gas emissions	P
THEME: INLAND WATERS		
Streams and rivers	River health (currently assessed using assemblages of macroinvertebrates)	C
Groundwater	Groundwater extraction	P
THEME: ESTUARIES AND THE SEAS		
Fisheries	Estimated fish stocks	C
	Total fish catch	P
Health of the marine environment	Extent of seagrass meadows	C
	Wastewater reuse	R
THEME: LAND		
Land use	Changes in land use	P

OBJECTIVES	TARGETS
Maintain or improve air quality in Adelaide air shed to meet NEPM or NHMRC guidelines	TSP target (maintain) TSP (lead) by 2000 PM-10 — inhalable by 2003 Sulphur dioxide by 2006 Ground level ozone (maintain) Nitrogen dioxide (maintain)
Return ozone levels to 1980 levels by 2020 to reduce amount of damaging radiation reaching the Earth's surface	Per cent phase out of HCFCs: 35 by by 2000, 65 by 2010, 90 by 2015, 99.5 by 2020, 100 by 2030 Per cent phase out of methyl bromide: 25 by 1999, 50 by 2001, 70 by 2003, 100 by 2005 Per cent phase out of halons (for essential uses): 100 by 2000
Reduce greenhouse gas emissions	108 per cent of 1990 baseline level by 2008-12 reporting period
Protect the ecological integrity of South Australian rivers and streams	All waterways to match or exceed criteria for moderate– good health
Ensure that water is available for current and potential future uses	Groundwater allocated should be less than or equal to recharge
Maintain fisheries populations for current and potential future uses	Maintain or improve populations of commercial and recreational fisheries
Manage fisheries for current and potential future uses	Catch matching resource capacity
Maintain marine habitat to sustain health of coastal environment	No further loss of seagrass meadows
Conserve biodiversity	
Prevent pollution of the marine environment	40 per cent or more of the discharges to Gulf St Vincent from WWTPs to be diverted to land-based use by 2001
Ensure optimal land use	No further alienation of prime agricultural land

Soil condition	Area of land affected by acidic soils	C
	Area of land affected by salinity	C
	Area of land with potential for wind and water erosion	P

THEME: BIODIVERSITY

Loss of biodiversity	Extent and condition of remnant terrestrial and marine vegetation	C
	Number of extinct, endangered and vulnerable species	C
	Distribution and abundance of key pest plants and animals	P
	Area held in protected areas eg. NPW reserves, MPAs	R

THEME: HUMAN SETTLEMENTS

Water consumption	Drinking water quality	C
	Water use per capita	P
Energy	Energy sources	P
	Energy use	P
Waste	Solid waste to landfill per capita	P
	Amount of material recycled	R
Protection of built heritage	Number of places on the State Heritage Register	C
Protection of Aboriginal heritage	Number of entities on the Register of Aboriginal Sites and Objects	C

Maintain productive capacity of land for current and future uses	No further increase in extent of acidic soils
Maintain productive capacity of land for current and future uses	No further increase in extent and severity of salinity
Maintain productive capacity of land for current and future uses	No further accelerated loss of soil due to wind and water erosion

Conserve biodiversity	No further loss or degradation of vegetation
Improve conservation and management of threatened ecosystems and species	No further extinctions or additions to lists of terrestrial threatened ecosystems and species
	'Downlisting' of current listed terrestrial species
Reduce impact of existing pest plants and animals	Control spread of existing pest plants and animals
Establish and manage a comprehensive, adequate and representative system of protected areas	Minimum of 15 per cent of each ecosystem held in national reserve system
	System of MPAs, including exclusion zones for habitat and species management, by 2003
	Minimum of 15 per cent of each marine ecosystem held in protected areas

Ensure drinking water meets water quality guidelines	100 per cent of water supplies comply with guidelines
Increase efficiency of water use	Reduce per capita consumption of water
Use alternative sources of energy to maintain current and potential future uses	Increase use of renewable energy sources to comply with Kyoto Protocol
Improve efficiency of resource use to maintain current and potential future uses	Increase efficiency of energy use to comply with Kyoto Protocol
Reduce resource use	50 per cent reduction by 2000 on a weight per capita basis from 1990 base
Reuse and recycle resources	Meet ANZECC targets by 2000
Identify, protect and conserve the State's heritage	Enter all places that meet the criteria in the Heritage Act on the Register by 2004
Identify, protect and conserve the State's Aboriginal heritage	Verify locations and report on site conditions for all sites on the Register of Sites and Objects by 2004

6

ECONOMIC MEASURES OF SUSTAINABILITY

TOR HUNDLOE

INTRODUCTION

There are 'defining events' in history, such as the invention of the steam train, the splitting of the atom, the discovery of penicillin. These and numerous other technological breakthroughs tend to come to mind rather than less immediately concrete events such as the formation of brand new institutions (such as the United Nations) or paradigm shifts in political philosophy (such as the vote for women).

The release of *Our Common Future* (alternatively known as the Brundtland Report) in 1987 is one of those defining events in philosophy[1]. As such it ushered in the concept of sustainable development: the notion of fixing up the world's major problems of poverty, food security, environmental degradation, depletion of non-renewal resources and, once fixed, ensuring that the biosphere is in good shape so that the needs of future generations are met. Two principles are involved: intra-generational equity and inter-generational equity. These two principles plus some others, which are more of an instrumental nature (such as the integration of ecology and economics in decision making at all levels), are radical changes in both political and economic philosophy.

As with other fundamental societal changes (for example, the emancipation of women), sustainable development has been more rapidly embraced in some countries than in others. By the date of the so-called Earth Summit in Rio de Janeiro in 1992 all nations were at least paying lip service to it. Even where acceptance of the concept was rapid, there is still some degree of discussion as to its meaning, and much discussion as to how to implement it. And then there is the issue of how to measure either progress or regress. This chapter focuses on this latter matter, with particular regard to how the discipline of economics and its tools can assist. However before commencing this task it is necessary to be clear on the meaning of both sustainable development and economics.

SUSTAINABLE DEVELOPMENT

Sustainable development progresses at two levels — the philosophical and the practical. Both are essential.

The word 'development' explicitly recognises that human wellbeing — health, living conditions, and social and economic conditions — can be, and for the poor must be, improved. Likewise, degraded rivers, oceans, forests, and the atmosphere can be rehabilitated. It makes sense to talk of improving the living conditions of other animals.

Those commentators who argue that the term sustainable development is an oxymoron miss the point made, first in the Brundtland Report and repeated by many since, that 'development' does not equate to 'growth' as traditionally defined. It is qualitatively different growth to that which GDP measures or that is implied in speeches by most politicians, business leaders and the financial media. What GDP measures will be discussed later.

Sustainable development is change for the better, and change that can be sustained. In terms of human wellbeing — if not necessarily in other contexts — it means change that is good, that is desirable according to commonly agreed ethical principles. This, of course, raises fundamental questions. But except for the moral relativists people can work towards identifying these. Only when these principles are agreed can means of measuring change be formulated.

The environment has fundamentally influenced human history at both an individual and societal level, just as humans have fundamentally altered the environment (to varying degrees depending on the period and the location). The concern in the 21st century is the extent of human influence (with a population of six billion and capable of doubling within 50 years) as well as the type of influence. Technological developments have given humans far greater scope than ever before to make very significant changes to the environment.

Take just one simple example of whether or not humans should interfere. The Great Barrier Reef is the largest coral reef in the world. It is of incalculable ecological value and a tourist icon of great importance to the Australian economy. At least twice in the past generation, large areas of corals were eaten by the Crown-of-Thorns starfish. There was considerable concern expressed by scientists, by tourist operators and the public. Should humans attempt to control and/or eliminate these starfish? If a population explosion was the result of some human interference such as the elimination of a predator, the answer would be 'yes'. If, on the other hand, this was a periodic, natural occurrence the answer would be that humans should not interfere.

Science can help answer such questions as this one (the reason for a population explosion). Questions like this are confronted continually when working to achieve sustainable development. The practical rule would be to do everything feasible to permit natural processes to be sustained. If that is the rule for the ecology, what rule should be applied to the economy?

Just as humans want a healthy environment, they also want a healthy economy. It is by pursuing what is known as economic activity that humans feed themselves, provide clothing and shelter, have jobs, make profits, generate the means of putting health-enhancing technologies in hospitals — and of putting smoke in the air, chemicals in the oceans, and drugs on the streets. The economy is people, and the machines, tractors, land, forests, fish and raw materials people work with. However, there are ecological constraints on the economy. The Brundtland Report and the existing vast literature on sustainable development recognise this fundamental point. If some choose not to recognise this, they are not talking about sustainable development. Those who criticise the sustainable development concept as being so imprecise and loose, that serious polluters can embrace the concept, are doing injustice to the idea. It is only by establishing criteria to measure progress towards a sustainable future that these differences of opinion can be put to rest. This chapter considers one approach, and not necessarily the most appropriate, to measure progress towards sustainability.

However, that is getting ahead of the discussion. We need to comprehend the parts of the sustainability equation better. The one we are dealing with here is the economy. Economies are dynamic. There is a need to understand much better how some economies develop, and why some falter or fail completely and people starve and die. There is no point in sustaining moribund economies. Consumer fashions (what economists call 'tastes') are continually changing. While this could be a result of successful advertising and promotion by leaders in the market, it does not mean that other producers do not have to respond, and they do.

Firms and economies that do not, or cannot, change what they produce, are not sustainable. We want to sustain economic processes — just as the goal is to sustain ecological processes — but with clear social end points in mind. These include eradication of poverty, sustainable jobs (or more appropriately sustainable incomes) and sustainable profits. Not only should firms meet the demands of the marketplace, but to do that within environmental and ethical limits. In other words, the economy is a human artefact and, in this sense, it differs from the environment. If we have the will we can change the way the economy functions, in terms of outcomes at least. Of course, we can take the view that we are relatively powerless to influence the economy, that 'the market', in particular the global market, is beyond human control. While we can take this view, it does not mean that it is at all accurate. In fact it is founded on a lack of knowledge of economic history.

There is a further consideration. This is the need to think about what sustainable development means in terms of sustaining social systems and sustaining cultures. Taking a long historical perspective, cultures and social systems are not, in general, static. They change; they evolve, but usually slowly. However, increasingly throughout the world (even in remote villages) modern communications technology is spreading messages, showing images, creating dreams that are hard to resist. The transistor radio, television and maybe one day the Internet, are having both desirable and undesirable impacts on cultures. These modern human artefacts are speeding up the rate of social and cultural change.

Clearly humans would want to sustain the good and the desirable aspects of a society and a culture. Learning from cultures different to their own enriches people. There are forces at work that lead people to better understand each other. By far the greatest gain from international tourism is the broadening of the mind. One can be optimistic that international trade and international tourism will be major forces in delivering a 'common future'. It is noteworthy that the title of the Brundtland Report, *Our Common Future*, is as powerful as the totality of the report's contents.

Just as there is hard work — research and scientific exploration — involved in understanding what is meant by, and what is needed to achieve, ecological and economic sustainability, so there is hard work in pursuing the development of ideas on which cultures and societies are based, be they religious or philosophical, and attempting to promote what is good and sustainable.

ECONOMICS IN A NUTSHELL

Economics is poorly understood by lay people, maybe more so than any other discipline. A major reason for this is that economists have given special, technical meaning to common words such as 'value' and 'efficiency'. In addition to that there is a tendency for the media to use 'economics' as an adjective when what is meant is 'financial', as will be explained.

Most of what is conveyed as economic information is in fact financial information. The profits and losses of businesses and the exchange value of currencies are obviously important financial data, but they are not necessarily economic data. The crucial difference between financial data and economic data is that the former is based on the individual or corporate perspective — taking as 'given' the whole array of taxes, subsidies, commercial practices and regulations that influence prices —

whereas the latter is concerned with society's economic welfare. Economists use the term 'social welfare' to define the *economic* state or wellbeing of a society.

The most significant correction needed to convert financial data into economic data would involve the removal of subsidies that *distort* investment decisions, and the inclusion of *externalities*. An externality is an economic cost or benefit that is not paid by a producer or consumer. The cost or benefit falls on someone else. A common example of an externality is pollution. A standard textbook case study is an upstream factory that uses a river to dispose of toxic effluent with the consequence that the downstream fishing industry suffers a major reduction in catches and profits. Externalities are pervasive in modern economies.

From a sustainable development perspective a fundamental principle is that a healthy environment is a prerequisite for a healthy economy. They are interconnected. This relationship has to underpin measures and indicators of sustainability. The question becomes is it possible to derive an indicator that is based on integration of the two disciplines of economics and ecology.

IS AN INTEGRATED INDICATOR FEASIBLE?

It is crucial to recognise that the conventional 'economic' indicator, gross national product or gross domestic product (GDP) is a very poor measure of income, economic welfare or anything else economic. It was not designed to be such a measure; rather the demands for data in semi-planned (capitalist) economies of the Second World War brought about a somewhat systematic gathering of statistics on production that took place *in the market*.

Defence personnel, national planners and treasury departments wanted to know how much defence output could be produced and what impact this would have on the whole economy. What happened outside of market was not a matter of interest. In 1945 (it is this recent), the League of Nations (the forerunner to the United Nations) convened a meeting on 'national accounting' and the world was started on the way to a universal form of national accounts known as the *System of National Accounts*. Only recently the United Nations (1993) established guidelines *Integrated Environmental and Economic Accounting* as a complement to the System of National Accounts. This initiative was in response to the 1992 United Nations Conference on Environment and Development (Agenda 21) and its sustainable development imperative.

Any properly trained economist recognises GDP for what it is — a reasonable measure of some things that happen in a country, and that might be important to that country. Daly and Cobb (1989, 64) make the point succinctly:

Indeed, no knowledgeable economist supposes that the GNP is a perfect measure of [economic] welfare. Most recognise both that the market activity that GNP measures has social costs that it ignores, and that it counts positively market activity devoted to countering these same social costs. Obviously GNP overstates [economic] welfare!

Anyone interested in the severe limitations of GNP, in attempts to make it a better measure of income, and in attempts to dramatically modify it so that it could be useful in measuring movements towards sustainability should consult Daly and Cobb (1989).

Attempts to improve GDP, such that it takes account of the contribution of unpaid work (an enormous amount is done in the household and in developing countries it is a fundamental part of the economy), the costs of pollution, the degradation of natural capital, the run-down of social capital and the distribution of income, have a long history (at least from 1972). The most recent attempts have been labelled the 'Genuine Progress Indicator' (GPI). The following seminal work is worthy of note: Nordhaus and Tobin (1972), Eisner (1985) Daly and Cobb (1989), Cobb and Cobb (1994) and Cobb et al. (1995). All these lead on to a number of country-specific formulations of GPI, including that by Hamilton (1998) for Australia.

To answer the question posed in the sub-heading — is an integrated indicator feasible? — it is necessary to return to the criteria for sustainable development. They necessitate that, at least, two equity criteria be met, intra-generational and inter-generational. With regard to the latter in particular, there is the clear need to understand the relationship between a healthy economy and a healthy environment, as they are interconnected.

The integration of ecology and economics *could* proceed by the conversion of ecological goods and services (such as fish-producing qualities of the oceans, rivers and coastal environments) into monetary terms. This is what economists attempt to do in social (extended) cost-benefit analysis of projects, and what they attempt to do in transforming GDP to an index of sustainable economic welfare (or GPI). This approach is making ecology part of the economy. The alternative approach is to make the economy part of the environment, and subject to the laws of ecology. The latter is what has to be done ultimately (as the laws of thermodynamics dictate). However, we need to ask if there is not a 'half-way house' in which the measurement of ecological goods and services in economic terms is not useful. Such measures could be worthwhile from two vantage points. Unpriced assets are too readily degraded; a common measuring rod overcomes some of the arguments about relative weights.

Notwithstanding the fact that GDP is a flawed measure of economic wellbeing, it has enormous political and public importance. The fact that it is flawed is not recognised by the public. Economists who know better use it rather than campaign to have it removed from the lexicon.

But even if GDP accurately measured changes in the growth of an economy, the issue remains as to the contribution economic growth makes to the overall happiness and wellbeing of people. Too many people assume they go hand-in-hand in a positive correlation. We will note that this is a flawed assumption, but press on with exploring how GDP can be transformed into something useful. Obviously by using GDP as the base we are considering the use of economic values for those things normally left outside of economics.

FROM GDP TO GPI

Heilbroner and Thurow (1998) write of looking down on the economy as from a plane or helicopter.

> to see it as a vast landscape populated by business firms, households, government agencies ... the ceaseless activity of production ... the never-ending creation and re-creation of the wealth by which the country replenishes and renews and expands its material life. This great central flow, on which we all depend, is called ... GDP. When TV newscasters say that GDP has gone up or down, what they mean is that the river of output has gotten larger or smaller, that we are producing more or less.

> As we look down on it ... There are hundreds of thousands, perhaps millions, of kinds of goods and services in the stream of production — foods of every conceivable kind, spectrums of clothing, catalogues of machinery, jumbles of junk ... this vast and variegated output can be divided into two basic sorts of production ... goods and services that will usually be bought by private households ... cars, haircuts, jewellery, meat, health care ... we call this branch of the river of production *consumption*, and the various goods and services in it *consumers' goods*.

> But ... we can see that there are also goods and services that *never end up in any consumer's possession* ... machines, roads, office buildings, bridges, airports ... office furniture and office typewriters ... We give them a special name — *investment goods* or *capital goods* ... they play a vital role in determining our economic wellbeing. To physical investment ... should be added outputs of educational skills in school and the knowledge produced by research and development ... *called human capital*.

What is crucial from a sustainable development perspective is that one stream (consumption) relies on the size and condition of the other stream (investment). Unless the latter is maintained the former will diminish over time. Some economists (for example, Daly and Cobb 1989) and Common (1995) argue 'that one way of interpreting sustainable development is to equate it with sustainable income, or what is known as Hicksian income' (after Sir John Hicks). Hicks (1948, 172) wrote:

The purpose of income calculations in practical affairs is to give people an indication of the amount which they can consume without impoverishing themselves. Following out this idea, it would seem that we ought to define a man's income as the maximum value which he can consume during a week, and still be as well off at the end of the week as he was at the beginning.

We can think of this proposition applying on a per capita basis in a nation and also globally. What is true for the week, is true for the year, the century, the millennium. Because the measure is on a per capita basis, while population continues to grow so will total consumption, and production will have to increase with the consequent increased demands on the environment. Therefore to obtain a realistic measure of sustainable income major adjustments have to be made to GDP.

Obviously depreciation of human-made capital (human-made machines, factories, offices) is a necessary step in moving from *gross* to *net* national product. This necessity is recognised in conventional accounting. It is done by imputation, which means the net measure is not as precise as the gross measure. Because the gross measure tells us something about market activity, and employment prospects, in the short term, it is the one that gains most attention by government and other decision makers.

The principle of applying depreciation to human-made capital needs to be extended to natural capital stocks, where natural capital are all those productive features of nature, such as soils, water, forests and carbon cycles. Of course, once this concept is accepted, on theoretical grounds there can be no argument as it is nothing more than a logical extension of what is done in conventional accounting, the difficulty is estimating values for natural capital.

Another correction is to subtract what economists term *defensive expenditure*. As the name implies, this is money outlayed to defend the community from unwanted side effects (externalities) of market activities. For example, if a new airport is built resulting in serious aircraft noise in homes and offices under the flight path, there will be (otherwise unnecessary) expenditure on sound-proofing. Another example could be where a new factory opens upstream and discharges effluent into the river that has to be then cleaned up by downstream users before it is safe, adding to the cost of production or consumption for these parties. Defensive expenditures are intermediate goods (like transport is in moving grain from farms to bakeries) that are a cost of production. Furthermore, it does not have to be production of something sold in the market; it can be production of the services of your home.

A further task is to add to GDP the monetary value of unpaid household work and voluntary work. Probably the best known, and most commonly stated, critique of GDP is 'if a man marries his house-keeper GDP declines'. Of course, it is not just work within the house but gardening, lawn mowing and so on, which is traditionally non-market activity. One of the drivers of increasing GDP in recent years has been the transfer of unpaid work in and around the house to the market sector. As a generalisation, urban residents no longer grow their own vegetables, keep their own laying hens, bake cakes or biscuits, make jams, and so on.

Because there are market prices for most housework chores (ironing, cleaning, babysitting, lawn mowing), it is possible to use these to estimate the amount of money that needs to be added to GDP. The principle is no different from the conventional approach of imputing a value for the home production of farmers and the rental value for one's own home.

In terms of dealing with what nature has provided 'free' it makes sense to think of two categories: natural resource (or natural capital) depletion, and environmental damage. Natural capital comes in two forms: renewable resources such as forests and fishing grounds; and non-renewable (except in geological time) resources such as minerals, particularly including fossil fuels. While humans can harvest the annual increment of timber or fish and meet the sustainability criteria, no level of use of non-renewables is possible if sustainability is imposed. There is, in theory at least, a way around this problem. It is to utilise the Hartwick rule, which can be summarised as follows. The sustainable rate of use of a non-renewable resource can be no greater than the rate at which a renewable source can be brought on stream to replace it. A fossil fuel could be deemed to be used sustainably if part of the profits from it were invested in a successful renewal energy source (say, solar, wind, tidal) that produced an equivalent stream of energy when the fossil fuel is gone. The means of handling this in converting GDP to GPI is to 'deduct an estimate of the amount that would need to be set aside in a perpetual income stream to compensate future generations for the loss of services from non-renewable energy resources' (Daly and Cobb 1989).

There are other important issues warranting general comment. These are the distribution of income and the value of leisure. As noted earlier, in conventional national accounting there is an implicit weighing of income: a dollar is worth the same to a poor person as a rich one. Proponents of GPI (and its predecessor) reject this on theoretical economic grounds and set about adjusting the money value of personal consumption in GDP to reflect the fact that a dollar increase to a poor person is worth more than a dollar increase to a wealthy person (see Daly and Cobb 1989).

The value of leisure is an interesting but frustrating concept to handle in national accounts. It is correct to consider leisure — the amount and quality thereof — as a positive contributor to human wellbeing; we can think of consuming leisure. However, to be precise in how much of anyone's non-working time is leisure rather than non-paid work (what is child-minding to a parent?) or time 'wasted' while under-employed or unemployed is near impossible. In addition to that is the matter of measuring the value of an hour of leisure. In cost-benefit analyses it is put at the opportunity cost, the wage rate; but is this appropriate across the population of a nation? For reasons such as these leisure was not included in the Daly and Cobb measure.

We should note that Hamilton (1998) in his estimation of the GPI for Australia adds in with the value of household work a sum for unpaid community work. Furthermore, he deducts amounts for un- and underemployment (the value of hours of idleness). These are legitimate adjustments but nearly as difficult to calculate as the value of leisure.

The net growth in human-made capital stocks is reflected in national accounts through the prices paid for final consumption goods and services. The use of the capital is part of the cost of goods and services that is recouped through sales. The same applies to pub-licly-owned capital such as power-stations where the monetary value of the services provided to consumers (say, electricity or water) are reflected in the national accounts by consumers' expenditure. However, there are government capital items such as roads that pro-vide services to consumers but are not charged for, or fully charged for, by governments and hence there is a need to account for these in the GPI.

As already noted, conventional national accounts are based on consumption, both by individuals and governments. These items are called 'private final consumption expenditure' and 'public final con-sumption expenditure', and are dealt with separately as the value of the services from public capital expenditure.

In Table 6.1 Hamilton's list is used, with a brief description of the item and how it is measured, plus where necessary the author's comment.

Table 6.1
Items added to and deleted from GDP
SOURCE Hamilton (1998)

Items in the GPI	How measured and comments
ADDED IN:	
Personal consumption: weighted	Personal expenditure weighted by an index to account for distribution
Public consumption	Subtracted from the expenditure by government on consumption are: (i) contributions to capital stock (which is included below); (ii) defensive expenditure. A major factor is determining the per centage of government expenditure made to offset a decline in wellbeing.
Household and community work	Hours worked in the household, less a percentage deemed to be 'non-marketable' activities, multiplied by the wage rate for housekeepers. Would seem to undervalue community work and some kinds of housework.
Services of public capital	The value of services from such capital investment as roads that are not charged for by governments and hence not reflected under consumption above.
Net capital growth	This is simply any net (after depreciation) increase in human-made capital stock, adjusted for population growth.
DEDUCTED:	
Costs of unemployment	Only the psychological costs are included, as other costs (reduced output, run-down of human capital and health impacts) are reflected elsewhere in esteem. How to value these costs in monetary terms is not obvious.
Costs of overwork	This is based on the concept that people work more than 'normal' hours on an involuntary basis. What they earn for these extra hours is one measure of this 'cost' to the economy.
Private defensive expenditure on health and education	Personal consumption expenditure on these two items is a positive amount in GDP. It is argued that a percentage of this is necessary to counter the health effects of pollution, and some education spending is defensive — in the sense that if one does not keep up with everyone else in the education system, one will fall behind in income earning ability (see Daly and Cobb 1989). Fairly arbitrary assumptions have been made in estimating GPIs.
Costs of commuting	A part of the cost of travel to work can be considered defensive expenditure. As a consequence of urban sprawl life-style is diminished unless households are moved further away form city centres. This is measured as the opportunity cost (wage rate) of travel.
Cost of noise pollution	An obvious case of defensive expenditure calculated from cost date.

Items in the GPI	How measured and comments
Costs of transport accidents	Costs of repairs plus costs of pain and suffering. How the latter are measured in practice is not clear. The theoretically correct measure is the amount of money a person would need to be compensated to experience pain and suffering!
Costs of industrial accidents	As for costs of accidents.
Costs of irrigation water use	The depletion of environmental flows is the impact to be valued. One measure of value is based on the value of water to irrigators multiplied by the quantity of water they will not have access to because flows are restored to an 'adequate' level.
Costs of urban water pollution	Pollution adversely impacts on downstream users and environments. One measure of the cost is the money spent on treatment; up to the stage that no further environmental damage occurs.
Cost of air pollution	Noxious air pollutants cause health problems for humans, other animals and plants. Some of the health costs to humans can be readily obtained. Other damage costs are much more difficult to estimate and in practice rough estimates are used. There is a need to be wary of double-counting health costs.
Costs of land degradation	The expected foregone output from land captures part of the costs of soil degradation. Estimates of this are available. In addition, there is some irreversible depletion of soil. It is a non-renewable resource. This would be accounted for by applying the Hartwick rule — see earlier discussion on 'natural capital'.
Costs of loss of old-growth forests	Forests provide an enormous range of ecosystems goods and services. Not all have been clearly identified yet and very few measured in monetary terms. However, there are some estimates of willingness to pay for preservation (from contingent valuation studies) and the results can be used as a crude estimate.
Costs of depletion of non-renewable energy resources	See the earlier discussion on natural capital.
Costs of climate change	This is an immense topic surrounded in much uncertainty as to the extent, timing and regional effects of the human-induced greenhouse effect. In the Australian GPI the cost is based on the expected price of purchasing a carbon dioxide permit to emit.
Costs of ozone depletion	In the Australian GPI, the estimated health costs plus a crude estimate of costs to agriculture and fisheries were used.
Cost of crime	This is one item for which it is necessary to be on the lookout for double-counting. There are good data available in Australia and most industrialised countries.
Net foreign lending	While foreign borrowing can contribute to sustainable income if invested wisely, this is not the case if it is used for consumption. It is the latter that is measured.

The various attempts to formulate more realistic measures of sustainability and progress by adjusting the conventional measure of GDP are worth pursuing and will continue to be as long as GDP is used by decision makers (and the public) as the reference point. Changing the measure does not guarantee sustainability but if it shows a decline we have serious cause for action. Both Daly and Cobb, and Hamilton, show a dramatic change occurring around 1970 with first a levelling-off then a decline in the late 1970s.

SATELLITE ACCOUNTS

Let us consider other approaches to measurement, but consistent with the theme of this chapter, ones that do not take us too far from economics. Throughout the 1990s there was a strong push towards 'environmental accounting' with environmental information gathered, analysed and published as 'satellite accounts'. As the name implies, these accounts are separate from, but attached to, the traditional economic accounts prepared under the international System of National Accounts. Satellite accounts are not alternative to, but a substitute for, integrated national accounting as with the GPI.

A small number of northern European countries have been preparing satellite environmental accounts. Only in the mid-1990s did Australia join the initiative. A small number have been, or are being, prepared, including a Water Account, Fish Account, Energy Account and a Minerals Account.

These accounts document the stocks and physical flow in terms of supply and use of resources (say, water) from the environment through various sectors of the economy. An advantage of this approach is that natural resource consumption can be correlated with economic activity. For example, the 1996–97 Water Account for Australia showed that industries (excluding utilities) generated a gross product of $382 000 per megalitre of water used, while the agriculture sector produced just $580 per megalitre. Within the irrigated agriculture sector, vegetables yielded a gross product per megalitre four times more than that of rice. Such information is potentially useful, for example, to policy makers considering the effect that water pricing reform may have on the economy.

One weakness of the satellite accounting approach, however, is that economic and environmental information must be collected on the same spatial boundaries in order for the analysis to be viable. Economic data are generally available only at a national (or sometimes state) level, but environmental data must be interpreted on the basis of catchments, drainage basins or biogeographical regions. For example, a megalitre of fresh water does not have the same environmental value in the wet tropics as in the semi-arid zone.

Satellite accounts report the consumption of natural resources, the impact on stocks of natural resources and give some indication whether or not the resources are being used sustainably. Until agreement is reached on how to prepare genuine integrated national accounts, satellite accounts will play a very important role in alerting society to the physical state of resources. They are a multi-criteria (at least two criteria) approach to considering sustainability. A somewhat similar approach (at a much grander scale) is the United Nations Development Programme multi-criteria measure of human development, the Human Development Index.

The Human Development Index (HDI) combines separate indicators of real purchasing power, education and human health. These are derived from more specific indicators, such as access to sanitation and access to safe water. The HDI is a multi-criteria indicator and, if it suffers from anything, it is the standard difficulty in weighting various criteria. Weighting is viewed as a subjective matter, and rightly so. However, it is important to be ever mindful of the fact that GDP implicitly treats a gain or a loss to a poor person as being commensurate with a same dollar gain or loss to a rich person. This is weighting of a particular kind and it is counter to the economic principle of diminishing marginal ability.

COST-BENEFIT ANALYSIS

So far we have been considering measurement at the macro scale. Everyday decision making is about the next project — to build or not build a hospital, a housing complex, a major water storage and hydro-electric station. It follows that each project should be subject to assessment in terms of its contribution (positive or negative) to a sustainable future.

Cost-benefit analysis has been the standard tool for appraisal of projects since the mid-1950s. It involves identifying alternative resource uses and calculating the costs and benefits (in dollars) of each. As concern for, and knowledge of, environmental impacts of projects grew, cost-benefit analysis evolved into what some call extended (or social) cost-benefit analysis. By this, they mean that negative and positive environmental impacts and (in some applications of the tool) distributional impacts (who gains, who looses) are included in the analysis (see Box 6.1). This is a major improvement of the early practice, and one clearly in concert with the original principles of applied welfare economics, which called for the inclusion of externalities and equity considerations in assessing projects. As outlined in this chapter, these are the very matters that underpin the modification of GDP such that it becomes a tool to measure sustainability. Many of the same problems with methodology associated with the 'genuine progress indicator' and

Box 6.1
Monetary values of the environment

The volume of literature on techniques for attaching monetary values to non-market environmental goods and services is now enormous. Some of the techniques are:

HEDONIC VALUATION

With this technique the value of a particular good or service is the sum of the values associated with its various attributes. Some of these attributes are related to the environment. For example, the value of a house may be the sum of values of attributes, which include size, age, proximity to services, the materials of which it is made, and (importantly for our purposes) its natural environs. Multiple regression analysis of the relationship between the price of a large number of (in this case) houses and these attributes allow a weighting to be assigned to each attribute.

CONTINGENT EVALUATION

Using this method people are asked how much they would be prepared to pay for an aspect of the environment if there was a market for it. The obvious drawback to this technique is that the people surveyed are not actually being asked to pay, so it is difficult to know if the responses accurately reflect what they really would pay in a real situation. However, practitioners have developed a range of techniques to design surveys and samples in ways that eliminate bias. An interesting feature of contingent evaluation is that the same group of respondents will nearly always indicate a higher willingness to pay to keep something they already have (for example, a national park) than to acquire exactly the same thing.

TRAVEL COSTS VALUATION

This assumes the value people place on the environment at a site is related to the costs that they are willing to bear in order to access it. A number of people are surveyed to discover how much it costs them to access the site and how often they go there. The value placed on the site is estimated by expressing the frequency of site use as a function of the cost of accessing it.

PRODUCTION FOREGONE

This estimates the value of goods and services provided by natural resources. For example, one of the environmental services provided by vegetation cover is to purify water in a catchment. If the vegetation was removed, it would be necessary to either build water purification plants to purify water for use by the people living in the catchment, or transport potable water from outside the catchment. The value of the 'water purification' service provided by the vegetation is equal to the cost of the purification plants (or on-going cost of transporting potable water, whichever is less). Possibly the most famous application of this approach was the 1997 publication in *Nature* by Costanza et al. of an estimate of 33 trillion dollars (US) for the worldwide value of 17 ecosystem services.

related measures also attend such approaches to cost-benefit analysis. For example, how can monetary values be meaningfully assigned to natural resources and how should adjustments be made to account for distributional impacts?

The need to treat the future as equal with the present is an additional fundamental difficulty in aligning cost-benefit analysis with the inter-generational principle of sustainable development. Cost-benefit analysis considers costs and benefits over a long period (at least the life of the project), and in summing costs and benefits over time, practitioners (taking their cue from human behaviour) are prone to assume that future costs and benefits are worth less than new ones. In other words, the future is discounted. Economists can quite adequately explain why this occurs in human decision making, and so far as 'conventional' assets (cars, cash, factories, overseas holidays) are concerned discounting is uncontroversial. However, there are vigorous debates over whether discounting should be applied to the environment (see Chapter 4).

There are two potential solutions to the problem that discounting causes in measuring a project's contribution to sustainability. One is for cost-benefit practitioners to, somewhat arbitrarily, set the social discount rate for the environment at zero. The other is to make it explicit that cost-benefit analysis is not the primary tool to determine whether or nor a proposed use of resources is sustainable or not. The protection of critical resources (natural capital) and the prohibition of actions that would cause irreversible impacts are matters that overrule any cost-benefit analysis. That is, it is only within the constraints set by the principles of sustainable development that we can apply the tool of cost-benefit analysis.

There are no 'fancy' rules to convert cost-benefit analysis into a universal tool to deal with sustainability issues. We need do nothing more than recognise and impose constraints. We take this approach with regard to a myriad of matters on which we make ethical judgements, such as child labour, the length of the working day and occupational health and safety. There is nothing new, or radical, in what economists call 'constrained optimality'. This is what Brundtland called for when arguing for 'growth' within environmental constraints. The idealists who criticise Brundtland should read what she says on this issue.

MULTI-CRITERIA ANALYSIS

Multi-criteria analysis is one attempt to place cost-benefit analysis within a broader decision-support framework that uses the strengths of cost-benefit analysis without forcing everything to be measured by economic criteria. Multi-criteria analysis begins by specifying the criteria for evaluating a project and attaching relative weights to each. Some of these criteria may be the costs and benefits dealt with in traditional cost-benefit analysis, but others may be couched in terms of job creation, regional development and the environment. It is important, that the factors relevant to each criterion can be expressed numerically.

As in cost-benefit analysis, a number of alternative resource uses are identified. For each alternative the factors relevant to the various criteria are quantified, then combined using the weights. The preferred alternative is thus identified. An obvious problem of multi-criteria analysis is that the relative weights assigned to the criteria are inevitably arrived at subjectively. Defenders point out that the technique at least makes these judgements explicit. Also it has the advantage that it allows an analysis of how much the result depends upon varying the weights.

CONCLUSION

Humans are condemned to make choices. The vast majority of people bring to bear a set of ethical principles in making choices. Notwithstanding an array of religions, racial differences and economic situations, there is ample evidence that humans can and do agree on some very basic principles. Most do not want war, starvation, poverty, crime and poor health. Humans are capable of empathy, something recognised by Adam Smith (the founder of economics) but not understood by those economists whose knowledge of human psychology is limited.

Resorting to underpinning the moral philosophy of neo-classical economics is just one approach to thinking about making choices. If we use this approach we need to know what the underlying ethical position is. We need to recognise, regardless of how we feel about it, that many of the most important decisions about the future are made by reference to GDP, which some think deals with human wellbeing. To apply the basic principles of economics rigorously to make GDP a better measure must be welcomed. However, as noted numerous times in this chapter, some things are above and beyond economics and there is nothing at all to stop us imposing constraints *before* we apply economic criteria in our decision making.

Our Common Future does chart the way forward in very broad terms. The concept of sustainable development that it presents to the world will be refined. Its critics either have not read it thoroughly enough, have little concern for the poverty in the third world, or are not practical people working for a better world. The final word shall go to Daly and Cobb, who gave us not only a much improved, environmentally-friendly measure of welfare, but also the outline of new, humanitarian economy. In referring to *Our Common Future* they state (p 371): 'As the concept of sustainable development is further defined, we believe it will begin to resemble our ... economics for community'.

NOTES

1   Of course, one should not view the arrival of the concept of sustainable development as a discrete event occurring at a precise time in history. The term 'sustainable development' can be traced back, at least, to the early 1980s while the concept, in a very general sense, can be found in ancient societies. It is a concept that has evolved through time, but it took the Brundtland Report to bring all the elements together.

7

SUSTAINABILITY INDICATORS: MEASURING PROGRESS TOWARDS SUSTAINABILITY

ANN HAMBLIN

MEASURING SUSTAINABILITY AT NATIONAL AND INTERNATIONAL SCALES

Sustainability, like beauty, seems to lie in the eye of the beholder. If we press the analogy a little further, we might say that in Australia the search for sustainability indicators has both its Ruskin-like idealists seeking integrated indicators of total sustainability, and the prosaic Philistines of the 'I may not know much about art but know what I like' school, intent on providing a corrective balance to an economics-dominated paradigm for specific sectors of the economy.

This latter sectorial approach has had the most active development of sustainability indicators to date in Australia. Implementation of sustainability indicators has been particularly associated with natural resource management, either in particular industries, such as agriculture, or within a region, such as the Murray-Darling Basin in Australia. In other sectors, such as manufacturing, health and transport the goals have often been set so broadly that realistic progress has been hard to develop or measure, while sustainability indicators that reflect total societal behaviour are still a long way in the future (Productivity Commission 1999).

Development and application of *integrated* indicators that have established relationships between environmental, social and economic attributes identified by Australia for internal self-evaluation is starting to occur (Eckersley 1998, Yencken and Wilkinson 2000). Internationally, Australia has also inevitably been included in many studies over the past decade that compare various aspects of sustainability and development across a large number of countries.

Most of these international comparisons stem from the United

Nations organisations, such as the World Bank, the United Nations Development Programme (UNDP), together with other multi-national groups such as the Organisation of Economic Co-operation and Development (OECD). Independent organisations such as the World Resources Institute (WRI) also publish comparative assessments of country performance based on combined indicators of economic, social and environmental development.

In 2000 the WRI and other organisations produced special global assessments of the world's utilisation of natural resources[1]. These ambitious publications have used 10 years of satellite observations as well as more conventional statistical information to provide spatial and temporal trends in resource use, so that eventually we may start to understand the *dynamic* nature of the relationships between population, natural resources and economic activities. The big step forward that this work represents is its capacity to be able to follow and interpret the dynamic, rather than static aspect of interactions between environmental, social and economic elements.

Nearly all country comparisons start from a basis of well-known economic indicators, such as the Gross National Product (GNP, which measures total output of residents and non-residents), or Gross Domestic Product (GDP, measuring total output plus net factor income from residents abroad) per capita. From these, and various population statistics, a suite of other social and socio-economic indicators is generally derived. Critics of the GDP as a measure of economic and social progress have been legion for decades. However, much of the blame for the abuse of the GDP must rest with central (finance) agencies of leading western economies and United Nations agencies that have encouraged the use of the GDP (and formerly GNP) to be used beyond its original intent (Waring 1988). This has inevitably led to the GDP becoming politicised, and used as an all-purpose index of the economy, or a shorthand for progress itself.

Furthermore, few published 'league tables' provide measures of the reliability of the statistics used as the basis to these indicators. Population numbers, for example, are notoriously unreliable in countries that have had long periods of internal disruption with breakdown in the normal census collection process, as in many African countries over the past 30 years, in south-east Asian countries during the 1970–80s, and in China during the 1960–70s. To check, supplement or provide data where none exists, the United Nations Population Division has a standard method of estimation of current country populations that uses fertility, mortality, and net migration data collected from sample surveys. However, the absence of coefficients of variation, or other error terms lends a spurious sense of reliability and authenticity to these 'league table' reports.

Despite this caveat, there are some intriguing and important indicators that have been developed for country sustainable development evaluation. The UNDP's Human Development Index, in particular, has made great intellectual strides in how to describe, monitor and interpret national trends in poverty, inequality and human rights (UNDP 1990–99).

THE HUMAN DEVELOPMENT INDEX

The Human Development Index (HDI) aims to measure the trends in human wellbeing across all countries. It was developed by those United Nations programs devoted to the alleviation of poverty and improvement in human condition, and published its first report in 1990. It is a compilation of three indicators that measure life expectancy at birth, educational attainment[2], and the standard of living measured as real GDP per capita. The individual indicators are converted to standardised scores, summed and averaged to provide the single HDI per country. The income index is a surrogate for all that makes up a decent standard of living, and is related to the money spent by each country, after debt repayment and defence expenditure have been subtracted from the GDP, on housing, water, food security, education and health. Aid and debt service ratios are provided separately. An example is provided in Table 7.1.

Table 7.1
Aid and debt services ratios provided as an adjunct to the HDI

Region	Net aid as % GNP		Debt as % exports	
	1991	1997	1985	1997
Least developed countries	13.2	11.1	20.5	12.4
Sub-Saharan Africa	12.3	6.7	25.2	13.7
South Asia	1.4	0.5	15.8	20.0
Latin America and Carribean	0.5	0.5	38.1	35.6
Eastern Europe and Commonwealth of Independent States	0.6	0.4	Nd	9.8

Source UNDP 1999

Beyond this simple ranking however, the HDI has expanded to consider many forms of intra- and inter-generational equity, including gender, age, rural–urban, and direct income distribution disparities. It has attempted assessment of human rights, law and order, the impact of new technologies such as the Internet, and industrial restructuring. In recent years the HDI has added short profiles of environmental degradation and managing the environment, but these are not integrated into the main index. The HDI has refined its methodology and data over a decade of operation, and sufficient aggregated data are available from the past 25 years for retrospective analysis to be undertaken on individual country progress over that time.

In its 'high human development' category, Australia comes out with flying colours, as having the fastest progress, from an index rating of 0.838 in 1975 to 0.922 in 1997, with Norway second and Canada third. For comparison, the fastest progress in the 'low human development' group of countries has been Indonesia (up to 1997), Egypt and Swaziland, while the slowest Burundi, Central African Republic and Zambia. Progress is a euphemistic term for this last category, where life expectancy has fallen by over ten per cent during the period. A total of ten African countries and eight countries in the Commonwealth of Independent States, (or former Soviet block) have experienced these tragic reductions in life expectancy through the combination of AIDS, economic collapse and civil disruption.

Unfortunately, the approach and methodology of the HDI has not been much used by the Organisation for Economic Co-operation and Development (OECD) in the work done in that organisation relating environmental with economic indicators. A recent set of OECD sustainable development indicator meetings (OECD 1999a, 1999b) recognises that popular issues in environmental concerns have driven indicator development, and that social indicators are lagging in their own right, and also in terms of their linkages and relationships with both economic and environmental issues. The approach being taken in the work by the WRI, previously described, may inspire the OECD working groups to look afresh at how such relationships may be achieved.

NATURAL CAPITAL, HUMAN CAPITAL AND THE GENUINE PROGRESS INDICATOR (GPI)

A number of influential thinkers such as Herman Daly, David Pearce and John Cobb, have been responsible for extensively challenging the use of the GDP from what might be called the inverse-viewing angle. That is, by considering those things that are costly to society (such as depletion of natural resources) as real costs, not simply as earnings (for example, Daly and Cobb 1989). This approach provided the intellectual basis for articulating the elements of sustainable development (World Commission on Environment and Development 1987). The thrust of these thinkers' work has been aimed at converting the intangibles of public good and public benefit into realities of monetary and non-monetary values. Central to this is the identification of the value of natural and human capital.

Natural stocks of resources used by all societies, such as water, air, land and vegetation, are distinguished from the flows of income and costs related to their use, and the value of environmental services that these stocks perform 'free' for society. Human capital evaluation takes into account not only the stock of people in society available to perform work and services, but also the costs (education, health services, crime prevention, and so on) necessary to maintain that workforce and the living standards needed to service a complex modern state.

One interesting development of indicators from this work has come via Redefining Progress, a non-partisan, non-profit research organisation devoted to developing an alternative to the GDP called the Genuine Progress Indicator. It broadens the conventional accounting framework to include family and community economic contributions and environmental values, differentiating between economic transactions that improve social and environmental wellbeing and those that diminish it.

Twenty indicators are used that include social costs such as crime and family breakdown, inequity of household income and defence expenditure (similar to the approach taken in the HDI), life span of material goods in the public and private sectors, and changes to work and leisure time. Environmental indicators focus primarily on the changes to resource stocks and their depletion (with implications for inter-generational equity), the hidden costs of pollution (such as deterioration of environmental stocks, costs to human capital), and the long-term effect of activities on all forms of capital caused by such agents as anthropogenic greenhouse gases, and nuclear wastes.

Examples of the GPI have focused on economic trends in the United States of America. The long-term trend in GPI suggests that while the American GDP has increased nearly linearly over the past 50 years from $US10 000 per capita in 1950 to $US30 000 in 1999, the GPI peaked in the 1970s (at about $US7000) and has since slowly declined to the same value as 1950 ($US5000). While some of the component indicators that are incorporated into the GPI are well substantiated, such as the growing gap between rich and poor in many Western societies, others are more difficult to assess, because the assumptions on which they are based are difficult to prove.

The long-term environmental effects of climate change are one such example. Climate change has become an accepted 'fact' in international circles, enshrined by the Intergovernmental Panel for Climate Change (IPCC), located within the United Nations Environment Programme (UNEP) and the World Meteorological Organisation (WMO). It has been given substantial international powers through conventions and protocols. Nevertheless, scientific predictions on the actual future climatic effects of rising temperatures and greenhouse gases are still tentative and are frequently modified, or even overturned by new evidence (IPCC 1995, 1998). Thus while the assumption that measured increases in greenhouse gas emissions will produce rising temperatures and atmospheric composition changes in future can be well substantiated, the impact of both these factors on *climates*, and *their* subsequent impacts on environments and human populations is still speculative. To assume that these climate changes will impact differentially on rich and poor, or negatively on the environment, as has been the case with some of the wilder claims both of non-government organisations and government institutions, only compounds the errors inherent in all scenario building.

Many of the attempts at producing 'triple bottom line'[3] composite accounts for sustainability purposes are regarded with caution, if not scepticism in conventional economic and financial institutions, because of the great uncertainties that surround the estimates of environmental accounts and services, and the impact of long-term forces such as climate change (Lutz 1993).

LAND USE, SUSTAINABILITY INDICATORS AND SOCIETY

Although agriculture contributes only a small proportion of GDP to most OECD countries, it still occupies the greater part of their settled land areas. In consequence the OECD has worked for over ten years on developing sets of measures to assess the interaction between agriculture and environment to achieve a better set of policy options and management outcomes.

Attempts to provide a consensus group of measures has been thwarted by the vested trade interests of individual countries in the fractious environment of their different positions on subsidies and tariffs. Thus the European Union (EU) block and Japan have had a strong emphasis on the significance of 'landscape values', whereas the Cairns group countries (particularly Australia and New Zealand) want more measures of economic performance. In 'New World' countries, agriculture tends to be viewed solely as an industry sector, rather than a cultural aspect of the society, as reflected in the detail and scope of provisional indicators (Table 7.2).

Table 7.2
OECD Working Group on Agriculture and Environment provisional indicators

Proposed indicator	Number of measures
Nutrient use	24
Pesticide use	12
Greenhouse gases	18
Water use	58
Land conservation	4
Biodiversity	43
Wildlife habitats	12
Landscape	38
Farm management	25
Farm financial measures	5
Socio-cultural issues	27

SOURCE Summarised from a restricted paper to OECD Working Group on Agriculture and Environment indicator development meeting, York, England, October 1998.

Of interest is the omission from this OECD group of indicators of any consideration of urban–rural relationships, either in terms of delivery of services in the relative economic value to the economy or to individuals. No analysis is provided or even suggested as to whether subsidies and support schemes are having a positive or negative effect on rural environmental quality, or retention of rural populations. This is all the more surprising as these issues are of significant interest to OECD agriculture and sustainable development working groups as policy options (OECD 1999a).

The Australian approach, exemplified in Case Study 1 described later in this chapter, reflects a utilitarian, production-focused agricultural sector, with indicators deliberately selected for an industry that is largely unsubsidised, in a free market economy. Long-settled OECD countries in contrast, wish to assess the sustainability of human-made agricultural landscapes and the inherited role of agriculture in long-settled societies where such landscapes are a large part of the national identity (Campbell 2000).

Some European and North American governments feel a degree of political unease in having economic indicators that reflect the actual financial performance of sectors which may be cross-subsidised within the overall economy. Australian primary industry sectors have received less than ten per cent subsidy assistance over the past two decades, compared with the average in the EU of over 45 per cent, and North American of 25 per cent over the same period. Australia, on the other hand, is challenged by the taunt that society is not prepared to pay such subsidies to restore and preserve agricultural environments, but continues to force its farmers to exploit natural resources for the sake of competitive exports on the world commodity markets.

The challenge in developing sustainability indicators for rural industries is to be able to demonstrate the reality of these differing situations so that consistent and appropriate policies can be developed. Internally, within Australia, the issue of cross-subsidisation between different sectors is not adequately reflected in sustainability indicators. Internationally, this translates into the relative values placed on sectoral economic, social and environmental performance in different countries.

A logical evolution of the OECD's work on agriculture and the environment would therefore be to assess the degree to which agricultural subsidies have a positive or negative impact on the environment in each country. The most recent meetings and papers from the OECD suggest, rather depressingly, however, that there is greater consensus to use the expanded System of Integrated Environmental and Economic Accounting as the basis of future work rather than draw on the equally well established, but less monetarily-focused HDI (OECD 1999a).

MOVING BEYOND THE CURRENT LIMITATIONS IN SUSTAINABILITY INDICATORS

CONCEPTUAL LIMITATIONS

League table comparisons that attempt to assess the environmental part of the sustainability equation have only recently moved beyond providing a suite of individual indicators, with countries ranked in relation to each. The World Resources Institute, Worldwatch Institute, and World Bank development reports have tended to follow similar systematic annual or regular reports on sets of indicators in separate categories of the environment. The new initiative coming from the Millenium Project of the WRI is a positive step in realising that *flows* (measured as rates of change) are as important as *stocks* (measured as units at an instant in time), in evaluating the sustainability of any society or ecosystem.

In the traditional international tabulated reports, categories are grouped under headings such as 'assets or stocks' and 'positive and negative impacts'. They include such general issues as atmospheric pollution, freshwater stress and availability, land use, deforestation, biodiversity protection, energy emissions and efficiency. However, as the World Bank (1999) report emphasises, the lack of a consistent framework used by all countries for environmental indicators makes it very difficult to compare across countries.

There are a number of frameworks that have been adopted. The best known frameworks are those that have been adopted by a number of countries, including Australia, in their Agenda 21 reporting:

- The Pressure-State-Response (PSR) framework is based on the idea that there are sets of anthropogenic pressures that affect the state of the natural environment, and that require societal responses to correct (Adriaanse 1993). This has a very strongly environmental focus, and suits government environmental responsibilities, but is unable to develop relationships between each segment or issue grouping. It also assumes that everyone is agreed on 'what the problem is'. This works well in certain cases, such as with single pollution issues (photochemical smog, toxic algal blooms), but not where there is dissension or ambivalence about the issue (household pets, culling kangaroos).

- Stocks and Flows Framework is based on the original concept of capital stocks (natural, human, built) and flows (movements from one stock to another) of classical economics. Social (human trust) and environmental (biomass) stocks can be added as separate components. Equivalent inflows to investment are resources, environmental processes and human honesty, and the commensurate outflows of depreciation are pollution, harvests and human corruption. It has a crude hierarchical structure, but no way of weighting the various elements. Recent developments in this field in Australia and internationally require sophisticated interactive models to simulate various sectors or issues (Cocks 1999).

- The Normative Framework, developed from the writings of Herman Daly (1973, 1988) extends these frameworks by relating natural wealth to ultimate human wellbeing through intermediate ends of health, shelter, education, and the built capital. The concept conceives of a broad base of ultimate means — the natural capital of ecosystems by which life is supported. Growing from this base are the built capital of our economic life and the human capital of social life. The ultimate end is the value system of individual self-respect and community self-identification, codified through law, religion and ethics. Sustainability indicators to this system identify whether people are well-off and happy, achieving this through least input of materials and energy, and maintaining natural systems in a functional manner for the long term. This is still the most ethically appealing framework to have been developed, but is beyond current information capture in the higher tiers. It is well suited to dealing with issues where there is no common consensus.

The selection of what constitutes a positive or negative condition may be a matter of debate, depending upon the degree to which the focus of the institution is on development or conservation. For example, agricultural production is seen as a vital aspect of development, particularly in countries with large and increasing populations, where reduction in the amount of agriculturally productive land per capita has an implied negative connotation. At the same time any loss of natural habitat for biodiversity is deplored, as is deforestation to provide more land for agriculture. The most appropriate framework in this type of situation will be one that allows for alternative pathways, as the Stocks and Flows or Normative frameworks do, rather than assuming a given consensus of opinion, as occurs in the PSR framework. It is also advisable to recognise the need for sectoral or issue-specific indicators to evaluate *trade-offs*. Trade-offs are not politically popular, but unfortunately they appear to be more common than *win–win* situations wherever natural resources are scarce and society's needs and wants are increasing.

LIMITATIONS THROUGH OVER-SIMPLIFICATION

In recent years the World Economic Forum, an independent grouping of American and Swiss research centres, has developed a pilot environmental sustainability index that attempts to construct a single measure of sustainability by combining 64 variables across five components into a single index (World Economic Forum 1999). The five components that have been selected are environmental systems, environmental stresses and risks, human vulnerability, social and institutional capacity, and global stewardship. The final resultant (environmental sustainability index) has then been plotted against the growth rate (GDP) and a 'competitiveness index', checked for correlations and the potential of these explored. At present this work is at a very early stage, and not very convincing; so many factors can combine in so many different ways to produce a similar index value. It is rather the same problem as a geneticist looking for the genes controlling yield in a crop — the genes for controlling disease resistance, or colour yes, but yield — well probably half of the total number may be involved.

There is a certain surreal air surrounding attempts to use single values for spatially heterogenous and discontinuous phenomena. An individual numerical value for an indicator of say, air or water quality, for a country the size of Australia, India or Russia, seems meaningless. So too do attempts at distinguishing the air quality of say, Luxembourg from that of Belgium, when the two are small, adjacent regions,

---

Box 7.1

The challenge of method and scale selection: the example of land clearing

Land clearing (removing native vegetation, cultivating land for agriculture) is an emotive and highly political issue world wide, nowhere more than in Australia.

- The concerns of those who advocate a total cessation are:

    –protection of biodiversity through retention of native habitats

    –maintenance of cover to minimise erosion and salinisation

    –preservation and enhancing of carbon sinks needed to fulfil Australia's Kyoto Protocol obligations.

- The concern of industry and government economic policy is the restriction this may have on commercial operations, and the dubious nature of some claims as to the benefits of non-clearance where land has already been much disturbed.

- Evidence of land clearing comes from three methods:

    –remote sensing of Landsat TM imagery (projected tree cover) change

    –surveys of farmers' actions (how much land have you cleared?)

    –numbers of licences or planning consents to clear being granted

- These actually measure different aspects of clearing: tree removal, land chained and ploughed, and intentions to clear (not necessarily carried out).

- The Australian Greenhouse Office needs to have a continental scale of reporting and to be able to partition above and below ground carbon removal and sequestration. Environment Australia's Biodiversity Section needs to know what vegetation is threatened by clearing, habitat by habitat. Natural resource managers need to know where deep-rooted vegetation and land cover is being removed in the landscape at catchment level. Economists want to know the costs and benefits of clearing versus retention.

Realistically these questions cannot be tracked using the same measurement, or even the same indicator (tonnes of carbon per year, area cleared per catchment, types of vegetation per biogeographical region, value of different land use options).

affected by the same air masses and have the same patterns of human activity. The far-reaching effects of the Chernobyl nuclear explosion will be sufficient to demonstrate how unrealistic such a distinction really is. Indeed, the whole concept of reporting on a country basis (that is using artificial, administrative boundaries) for natural systems that have their own physical boundaries, may provide built-in potential for such serious error, that the resultant values may be very badly misused.

This view has been strongly endorsed recently by the expert group working on sustainable development frameworks for the OECD:

> ... the point of departure ... is the recognition that no single number can encapsulate all of the relevant information for sustainable development policy formation. (OECD 1999).

At the same time, member countries are striving to develop a core set of indicators that can encapsulate the issues of sustainable development for public information, awareness and behavioural change. As a result the United Kingdom has been keenly promoting a suite of 'headline indicators' for this purpose, as have Sweden and Australia (refer later section in this chapter).

Inevitably individual needs and sustainability indicators required at local level are not well catered for by broad-brush generic indicators. Box 7.1 provides a topical Australian example of why we can get different results from asking slightly different questions about the same topic — and at different scales.

CASE STUDY IN CHINA: OVERSIMPLIFICATION OF CROPLAND PER CAPITA AS AN INDICATOR OF FOOD PRODUCTION

'Cropland per capita' or 'population per square kilometre of arable land' is a frequently used indicator of agricultural production capacity. Simply dividing the number of hectares of agricultural land by the population however, gives no indication of the capacity of a country to feed itself. This is vividly borne out in the excellent recent in-depth analysis of China's situation by Gerhard Heilig of the International Institute for Applied Systems Analysis, Austria (Heilig 1999).

This study clearly demonstrated how reliance on poor estimates of population and cropland such as those used by Lester Brown (1995) in his provocative book *Can China feed itself?* led to Brown's misinterpretation of the relationships between population and food production in China, and to his conclusion that the country was already facing a future food crisis.

Heilig's analysis (Heilig 1999) shows that the estimates of China's projected population growth to 2025 cannot be more accurately determined than to + or - 200 million (in 1.38 billion). His work shows that changes in the proportion of crops grown for animal consumption (now 40 per cent), conversion of cropland to orchards, fish-ponds, and other types of diversification, and statistical under-reporting of agricultural land have all contributed to a 22 per cent under-estimate in the actual amount of agriculturally productive land in China. Food production in China is not simply a matter of growing rice and wheat for starving masses anymore. It is a complex issue of food industries, government policy adjustments to market development, increasing consumer preferences and rising standards of living. In other words, sustainable development involves increasing the quality, not just the quantity of life.

The very important lesson from Heilig's study, for all those involved with sustainability indicators, is his demonstration of the limitations of the population statistics, and of the gross under-estimation of China's croplands from the official statistics, provided by provincial administrative returns. China Statistical Yearbook in 1998 showed an official estimate for cropland of 95 million hectares, whereas the collaborative remote-sensed satellite data collected by the US-MEDEA group and China State Land Administration have shown the real figure is 131.1 million hectares. While some of this under-reporting arose from the reliance on a system of agricultural statistical collection that tied productivity gains to provincial funding from central government, some arose from the simple lack of a set of descriptors for land in anything other than cereal production.

SUSTAINABILITY INDICATORS FOR NATIONAL POLICIES

NATIONAL HEADLINE INDICATORS

Work to develop a set of national headline indicators for Australia has been developed by the ANZECC State of the Environment Task Force and elaborated at a recent Commonwealth workshop (Australian Bureau of Statistics 2000). 'Headline' indicators were described as those that are able to report on the key issues of the National ESD Strategy (see Table 7.3). Government departments proposed set of 22 indicators, but appear to have struggled with how to reflect inter-generational equity, although intra-generational equity is represented, albeit modestly, by a few of the indicator suite already used in the HDI series (UNDP 1990–99). Perhaps this is not surprising given the lead agency's perspective in which environmental issues are at the forefront, and given the very recent recognition in many OECD countries of the need to have more direct articulation of the social dimension of ESD in their indicators.

One of the most valuable indicators used as part of the HDI, and later adopted by other UN agencies and independent study groups, has been the use of the Gini Coefficient or Index. This relatively simple measure of the inequality of income distribution, so useful in time trends on intra-generational equity, does not rate a mention in the

Table 7.3
Preliminary headline indicators for Australia

SOURCE Australian Bureau of Statistics 2000

Aspect of Core ESD Objective	Headline Indicator

CORE OBJECTIVE 1: INDIVIDUAL AND COMMUNITY WELLBEING AND WELFARE

Living standards, economic wellbeing	1 Real GDP/capita
	2 Gross household disposable income
Education and skills	3 Per cent people attaining secondary and tertiary qualification
Healthy living	4 Average expected years of healthy life
Drinking water	5 % population with access to drinking water by treatment and quality
Air quality	6 Exceedences of NEPM standards for major urban area ambient air quality
	7 Total SOx, NOx, and particulates

CORE OBJECTIVE 2: ECONOMIC DEVELOPMENT THAT SAFEGUARDS WELFARE OF FUTURE GENERATIONS

Economic capacity	8 Multi-factor productivity
Industry performance	See 1 — Real GDP/capita (chain volume measure)
Economic security	9 National net worth
Use of natural resources — Eco-efficiency	
— water	10 Water extraction as a proportion of extractable yield
	11 Water use/unit GDP
— forests	12 Extent of area by forest type and tenure
— fish	13 Status of commercially viable fish stocks
— energy	14 Total non-renewable energy use and per GDP
— agricultural and pastoral land	15 Area of agricultural and pastoral land affected by land degradation
	16 Catchment condition index

CORE OBJECTIVE 3. INTER- AND INTRA-GENERATIONAL EQUITY

| Intra-generational equity | 17 Distributional information by gender, age, health status, ethnic origin, location, income, occupation for: Living standards and economic wellbeing, Education and skills, Healthy living, Drinking water quality |
| Inter-generational equity | Only indicator proposed is the collective results of all indicators; if trends show positive values in these then inter-generational equity is probable: — individual and community wellbeing enhanced, intra-generational equity provided for, biological diversity, processes and life support systems being maintained, economic development maintained and improving. |

CORE OBJECTIVE 4: PROTECTING BIODIVERSITY AND MAINTAINING ECOLOGICAL PROCESSES

Biodiversity and ecological integrity	18 Extent and condition of representative terrestrial, coastal, freshwater and marine ecosystems, including extent of reserve and non-reserve representation
	19 Number of extinct, endangered and vulnerable species and communities
Climate change	20 Total CO_2 equivalent emissions, per GDP
Coastal and marine health	21 Extent of marine disturbance (interim)
Freshwater health	22 Macro-invertebrate assemblages
Land health	23 Covered by indicators 10, 15, 16, 18, 21, 22

current sets of core indicators being canvassed in the OECD countries. This is curious, when one of the concerns expressed widely in social commentaries in post-industrial countries during the past decade has been that income distribution is tending to become bi-modal, or polarised, despite the overall increase in total wealth globally, and per country.

Around the world many groups, including the Australian 'headline indicators' task force, are struggling to find better ways of expressing human aspirations (Daly's ultimate ends, the ethical outcome), which will provide indicators of social cohesion, trust, civil liberty and legal justice. This is an acknowledged difficulty and gap in all such work, yet to date it has been easier to pick on indicators of alienation (drug abuse, rates of violence, suicide) than on the positive values expressed by free speech, the rule of law and the right to vote. OECD is conducting work in this area, which is still ongoing. Hopefully this will be the basis to the next evolution of headline indicators.

INDICATOR SELECTION FOR ESD IMPLEMENTATION

Australia's involvement with sustainability stems from environmental concerns with prevailing economic theories of the late 1980s, and the Commonwealth Government's adoption of the principles of ecologically sustainable development (ESD) in 1990.

Australia's definition of ESD was couched in consciously environmental terms:

> ... using, conserving and enhancing the community's resources so that ecological processes, on which life depends, are maintained, and the total quality of life, now and in the future, can be increased (Commonwealth of Australia 1990).

It drew heavily on the definition of the World Commission on Environment and Development (1987) of 'meeting the needs of today, without compromising the needs of future generations'. A National Strategy for ESD (NSESD) was published in 1992 (Commonwealth of Australia 1992). It was endorsed by all Australian governments in the same year, and sought to integrate long-term with short-term societal goals with a very broad policy agenda reaching into all parts of government and society's activities.

The strategy was made up of a set of sectoral and inter-sectoral issues, with some generalised pointers for implementation. It is worth considering what progress has been made in the past eight years in the sectoral issues. Table 7.4 draws on government legislation, Industry Commission (since 1998 Productivity Commission) reports, and sectoral association annual reports to compile a snap-shot of the types of changes that have taken place in Australia in the past eight years.

Intersectoral issues that formed the major portion of the NSESD

Table 7.4
Summary of progress towards ESD in sectoral issues

Sectoral issues	Government policies
Agriculture	Natural resource programs, taxation relief on landcare, business and technical training. Establishment of National Registration Authority (chemicals), Natural Heritage Trust grants program for environmental management.
Fisheries	Adoption of a fisheries ecosystem management approach for all commercial fisheries, development of aquaculture industry, strong Australian pressure for improved international fishing standards and implementation of the Law of the Sea, incorporation of ESD into all fisheries legislation and a National Bycatch Policy.
Forests	Regional Forest agreements for state forests (1997–2000), sustainable management codes of practice developed by 1995 onward in most States, with government developing a nationally agreed approach to Forestry Standards that will assist certification of products to specified standards.
Manufacture	Increased policing of waste management, pollution control (emissions) and energy efficiency via state EPAs and other state regulatory bodies established. Occupational health and safety requirements for greater staff training, upgrading of facilities.
Mining	State guidelines for mine site rehabilitation and management following ANZECC/NHMRC guidelines (1992). Australian Minerals Council devolves operational management of environmental monitoring, implementation of standards and regulatory compliance to industry.
Urban	Better Cities Program leading to attempts at city infill, re-development of twilight zones, amalgamation of local councils in response to increasing service delivery demands, early adoption of Agenda 21 strategies by local governments with strong commitment to recycling and waste management programs.
Tourism	Eco-tourism focus, with major tourist destinations mostly World Heritage Listings. Efforts to manage environmental impacts with a National Plan (1998) for sustainable develop-ment, and response from state and Commonwealth govern-ments to provide adequate planning, regulatory and financial support to growing industry.
Energy	Kyoto Protocol with establishment of National Greenhouse Office 1998, energy reduction programs, but significant continuing government support (licensing, incentives) to fossil fuel exploitation, with less than three per cent investment into alternative power generation sources.

Industry responses

Quality accreditation schemes, increase in private consultants servicing agriculture, patchy natural resource responses, increasing polarisation between profitable and unprofitable farms, with majority of smaller farms unable to afford environmental remediation.

Establishment of Management Advisory Committee with broad stakeholder representation, use of logbooks for recording catch and effort, regular assessment of status of stocks, increased licensing stringency, industry structure to reduce effort, catch restrictions, industry codes of conduct. Increased effort to reach agreement on high seas and prevent illegal harvesting.

Private forests variable management, voluntary compliance from large companies but lack of data and reporting from small private owners. Rapid increase in plantation investment, particularly in hardwoods stimulated by carbon credits, and sophisticated investment schemes in southern Australia.

Cradle-to-grave waste management systems, developed along with energy efficiency systems, self-auditing of performance, particularly in industrial chemicals and energy companies. Smaller operators struggle with costs, and sophistication of requirements.

Industry voluntary adoption of best management guidelines, and independent Australian Minerals and Energy Environment Foundation (since 1991). All major companies employing environmental staff with community involvement and publication of annual reports on environmental monitoring, rehabilitation, risk and waste management.

Very different capacity of rural and urban councils to deliver services, with increasing polarisation, investment in coastal regions and major transport routes. More articulation of conflicts of interest between conservation and development, and strong growth in privatised and outsourced services.

Rapid growth in the sector has seen tourism grow to nearly six per cent GDP, and has led to strong interaction with government and non-government organisations. Industry associations and environment groups (for example, Wilderness Society) working together on environmental planning for major tourist destinations, tour operation, facilities and promotions.

Australian Petroleum and Mineral Guidelines: voluntary company adoption of comprehensive guidelines for best practice management of all stages of petroleum, gas and coal exploration, mining, refining, transportation, distribution, sale and disposal of wastes. Establishment of industry supported foundations and research centres.

had a strong environmental thrust. This was understandable in the light of the enormous focus on redressing the under-representation of environmental issues, measurement and monitoring brought to world attention by the United Nations Conference on Environment and Development (UNCED) at Rio de Janiero in 1992.

Of the 21 issues listed, ten are devoted to issues of environmental conservation, management, research and information, although the NSESD policy has a strong anthropocentric (rather than teleo-centric) focus:

- enhance individual and community wellbeing and welfare by follow ing a path of economic development that safeguards the welfare of future generations; and

- provide for equity between and within generations.

Equal weight is presumed to be needed among economic, social and environmental issues, but economic issues were curiously under-repre-sented in Australia's NSESD (Table 7.5), particularly in relation to intra- and inter-generational equity.

Table 7.5
Economic, social and environmental inter-sectoral issues of Australia's NSESD

SOURCE Commonwealth of Australia 1992

Economic issues	Social issues	Environmental Issues	Integrative issues
Pricing and taxation	Aboriginal, Torres Strait Islanders	Biological diversity	Land use planning and decision making
Industry, trade and environmental policy	Gender issues	Nature conservation system	Changes to government institutions and machinery
Employment and adjustment	Public health	Native vegetation	International co-operation, overseas assistance policy
	Occupational health and safety	Environmental protection	Research, development and demonstration
	Education and training	Natural resource and environmental information	
	Population issues	Environmental impact assessment	
		Coastal zone management	
		Water resource management	
		Waste minimisation and management	

Environmental issues are strongly emphasised in the 1992 National Strategy because these were the areas that had been largely overlooked in the historical development of Systems of National Accounts (SNA). At the end of the 1980s, with economic rationalism strongly in the saddle, the greatest challenge to economists and statisticians was to develop supplementary systems of national accounts that could adequately value natural resource assets. Marilyn Waring (1988) has vividly described some of the struggles and challenges faced by international working groups trying to come to grips with valuation of such things as unpaid labour, ecosystem services and public goods. In conventional SNA methodology, anachronisms abound; such as a tree only having value once it is cut down and sold, or the GDP being reduced when a widower marries his housekeeper.

More recently supplementary national accounts and global valuation of ecosystem services have been attempted (Pearce and Anderson 1993, Australian Bureau of Statistics 1998, 1999), but by the very nature of their current selection of measures, they tend to state the obvious. Countries with large land areas, and plentiful natural resources coupled with small populations (such as Australia and Canada) show up well alongside those with high population densities and small natural asset base (such as Japan and the Netherlands).

If the NSESD had been written in 2000, it is likely that greater attention would be focused on some of the issues that have become politicised in the past few years, such as differences in urban-rural economic opportunity, service delivery and communications. Concern with national economic security has probably also increased since the period of precipitous and un-forshadowed collapse in the Asian economies in the mid-1990s.

These changes in emphasis suggest that, while governments have supported and tackled sustainability issues in specific sectors, they have had greater difficulty in tackling inter-sectoral actions, and supporting the view that society's interpretation of what is meant by sustainability is still a moving target.

Governments and commercial sectors have had strong vested and legislative interests in adopting ESD at a corporate, planning and policy level. Strategic plans, guidelines, policies and principles abound in both public and private sector documentation, and are now a standard part of corporate reporting in the private sector. However, progress in bringing about actual changes to practice, and devising ways to measure such changes through sustainability indicators has inevitably been slower and patchier. An investigation of the progress in adopting ESD principles and practices in the federal public sector found that many government departments do not have clearly articulated policies and programs in place to bring about 'adaptive management' for sustainable development (Productivity Commission 1999). Frequently the paradox occurs of counter-indicative policies and programs, as with one part of a government department encouraging alternative energy sources, while another part is promoting development and growth in fossil fuels.

Actual *use* of sustainability indicators in Australia has been restricted to some industry sectors. The following case studies show how iterative, complex and time consuming the process has to be if the results are to be accepted by the stakeholders as meaningful, and used in an effective manner.

AUSTRALIAN CASE STUDIES

AUSTRALIAN INDICATORS OF SUSTAINABLE AGRICULTURE

In 1991 Standing Committee on Agriculture prepared a report on sustainable agriculture (Australian Agricultural Council 1991), which endorsed a program of work to improve the sector's sustainability. This program included the development of indicators of sustainability (Agricultural Council of Australia and New Zealand 1993) and from these proposed indicators, a nationally co-ordinated study to track the progress of agriculture over the period 1986–96. The indicators were developed with the object of providing the agricultural sector with a means of assessing sustainability against economic, environmental and social criteria, both in terms of 'internal' performance, and 'external' performance. A minimal set of indicators was selected that would be applicable across all industries, and regions, with each indicator being interpreted through a number of attributes (Table 7.6).

A pilot scheme tested the proposed indicators and attributes in some regions and industries, followed by substantial revision of the final, agreed attributes. Methods were documented, with justifications, limitations and data sources described in the final report (ARMCANZ 1998).

The project involved over 100 public and private sector researchers and statisticians across Commonwealth, state and territory jurisdictions, and took three years to complete. It provided a set of situation statistics, some trends in individual attributes, and has been the basis to further federal government initiatives in this area. The study has been widely consulted in the OECD Agriculture and Environment working groups, where the selection of indicators chosen has generated considerable discussion.

The report has focused additional political and administrative attention on addressing issues, such as the low financial performance and specific types of environmental damage. It was not able to assist in providing an integrated 'score card' of which regions or industry sectors are more or less sustainable *as a whole*. Integration, thresholds, trade-offs, and targets and challenges for the next reporting phase for this set of indicators, is planned for 2001–2, after the next national population and agricultural censuses. In the meantime, it is expected that the National Land and Water Resources Audit, currently being undertaken across Australia, will add enormously to the reliable and topical information on natural resources that can be used (Commonwealth of Australia 1997).

Table 7.6
Australian indicators and their attributes, for sustainable agriculture
SOURCE ARMCANZ 1998

Indicators	On-site (internal performance)	Off-site (external performance)
Economic indicators and their attributes	Long-term real net farm income • real net farm income • average real net income • farmers' terms of trade • debt servicing ratio • total factor productivity	None developed Externalities that affect agriculture are already monitored through general economic indicators such as exchange rate and credit rating
Environmental indicators and their attributes	Natural resource condition • water use by vegetation • soil acidity, sodicity • nutrient balance • rangeland condition • agricultural plant species diversity	Off-farm environmental impact of agriculture • chemical residues in products • impact of agriculture on native • vegetation • dust storm index • salinity in streams
Social indicators and their attributes	Managerial skills • level of farmer education • extent of participation in training and landcare • implementation of sustainable practices	Off-site socio-economic impacts • age structure of agricultural workforce • access to key services

The same type of framework has been more readily used at regional or state level, where detailed information is available for adjusting policies to local community demands. The Murray-Darling Basin Commission for example, has brought together a very large body of information on water flows, a salinity audit and regional ecological, economic and social needs to provide a basin-wide strategy for combating rising salinity (Murray-Darling Basin Ministerial Council 1999).

Finally, the question should be asked, who uses this report? Have the farming community themselves made use of it? The answer is, on the whole, no. Do their industry peak bodies use it? Not to any great extent. What about governments? State and territory agricultural agencies have based their policies and programs on the SCARM publications on sustainable agriculture, and some have adopted the same indicators, with the same reporting framework. The OECD has been interested, and has had a continuing dialogue with Australian Commonwealth departments on the indicators used. In the final analysis, we can conclude that this exercise has synthesised many disparate studies and confirmed views on the relatively strong production performance, and poor environmental, economic and social condition of Australia's agricultural sector.

REGIONAL FOREST AGREEMENTS AND MONTREAL PROCESS INDICATORS

Australian public forest management reached a crisis point by the late 1980s, when conservation groups staged repeated confrontations with forest industries in one region after another. As in North American west coast forests, and those of tropical rainforests in many countries, the issues were the perceived destruction of forest biodiversity, and in particular the plight of icon species whose survival depended on big old trees being retained in 'old growth' forests.

By the early 1990s this community pressure widened both to include general community concern about the state of forests, and to acknowledge the full range of biological diversity that should be preserved. The NSESD and the ratification for the Convention of Biological Diversity in 1993 stimulated joint government and forestry industry working groups to develop ESD management plans and guidelines for publicly owned forests. It is important to keep in mind however, that almost 70 per cent of native forests in Australia are on land managed by the private sector. While all forests are controlled by some level of policy and management regime these are very variable on tenures other than the public multiple-use tenures and conservation reserves (National Forest Inventory 1998). In the forestry policy statement the states, territories and Commonwealth agreed to a broad spectrum of development, production, conservation, recreation and management goals (Commonwealth of Australia 1992a).

Comprehensive regional assessments have since been carried out on parts of the forest estate as a basis to negotiations on agreements that set out how forests are to be managed over the next 20 years. These Regional Forest Agreements (RFAs) will be implemented through state legislation and policy mechanisms. They have been the subjects of substantial discussion and negotiation, and nine were signed by the end of March 2000.

How will we know whether the management of these forests is being carried out in what is considered to be a sustainable manner? Each RFA contains regular public reporting requirements against agreed indicators that should show progress towards sustainable development. For regions where commercial forestry is a major forest use, RFAs have assessed the full range of forest values, more comprehensively than has been done for any other extensive natural resource in Australia. As a result, there have been improvements in forest management practices, and in the extent of forests preserved for biodiversity protection. The explicit attention to social values was the first major attempt to involve communities in what has often been a two-way debate between environmentalists and economic interest groups. The RFAs provide a sound basis for on-going management of all forest values, but the RFAs are not static. There is still room for improvement in many areas and the agreements are expected to be responsive to new findings and changes in economic, social and environmental information.

Sustainable forest management is a complex issue, which involves managing forests for a wide variety of uses, including commercial production, biodiversity and water conservation, recreation and maintenance of traditional (indigenous) gathering and shelter activities, and most recently, use of forests for carbon sequestration to mitigate greenhouse gas emissions. This implies there may be trade-offs, and potential for continuing argument over resource uses that conflict in their outcomes, though not in their aims.

Considerable reliance is being placed on the development of indicators through the Montreal Process, to which Australia became a signatory in 1995. The Process involves 12 countries that together represent most of the world's temperate and boreal forests, and some tropical forests. They have developed a comprehensive framework, known as the 'Santiago Declaration' of 67 indicators relating to seven criteria (Canadian Forest Services 1995):

- biological diversity

- productive capacity

- ecosystem health and vitality

- soil and water resources

- global carbon cycles

- long-term multiple socio-economic benefits

- effectiveness of legal, institutional and economic framework

Most State and Territory forest agencies have changed their mode of organisation during the 1990s as a result of the continued community pressure for greater transparency and accountability of their revenue-generating operations. However, the impact of these changes on actual forest condition and silvicultural systems is still patchy, and does not necessarily extend to other tenures, such as private forests, or the very substantial areas (66 million hectares, or nearly 40 per cent of the total forest estate) that are in leasehold tenure. In the latter the main land use is predominantly sheep and cattle grazing, with forests being thinned to promote understorey vegetation for livestock grazing.

In the case of the forest sector therefore, a very substantial body of knowledge is being collected and curated within Commonwealth, state and territory information agencies, to support a comprehensive indicator system. The acceptance of these tools to implement ESD has been a contentious political process, with stakeholder acceptance across the whole forest estate.

A NATIONAL ESD REPORTING FRAMEWORK FOR FISHERIES

The Standing Committee on Fisheries and Aquaculture has recently embarked on a major project to develop a national reporting framework for ESD. This action has been prompted by the increasing demands of other government regulatory agencies, non-government organisations, consumers and the general public for evidence that fisheries are being managed according to principles of ESD. The project involves several steps including an agreed terminology, development of high level objectives, a series of case studies to test the approach and subsequent application of the reporting framework to each of the approximately 140 managed fisheries in Australia (Whitworth et al. 2000).

The reporting framework follows a similar approach to that developed by the Bureau of Rural Sciences (Chesson and Clayton 1998) and is consistent with FAO Technical Guidelines (Garcia et al. 1999). The reporting unit is an individual fishery, the same unit with which management decisions are made and implemented. The framework seeks to identify the fishery's contributions to ESD (both positive and negative) across all aspects — environmental, social and economic. The reporting structure is adapted to each fishery through a consultative approach, involving all stakeholders, and the focus is on specifying operational objectives against which progress can be measured. Indicators then follow *after* objectives have been specified, so that a direct link is made between performance, as revealed by the indicator, and management responses.

LESSONS LEARNT

The lesson learnt from these case studies suggest that extensive community and stakeholder consultation is necessary throughout the process of development of sustainability indicators intended for management application. Forestry and fisheries sectors have used this approach to a greater extent than agriculture, where the process has been largely one of government administrative and research agencies' activities. The forestry process has resulted in a substantial change in the way forests are being managed in Australia, and the ESD framework for fisheries has opportunities for similar positive benefits. In the case of agriculture, a much broader suite of activities has been occurring in parallel with the indicators process, several of which are also the result of government initiatives, such as the national Landcare program, business training programs, and industry restructuring packages.

Another lesson learnt is that these consultative processes increase the cost and time taken, and may not be needed if the objective is to provide indicators for policy planning or development. Nevertheless, the process still needs to be transparent, and engage stakeholders in agreed sets of objectives if the resulting indicator suite is to have acceptance. It may be wishful thinking to expect sustainability indicators to serve both as effective management tools and as informers of public policy. The reasons may well be found in the mundane technical issue of scale.

Probably over 90 per cent of all data used in ESD indicator development has some component of spatial and temporal scale attribution associate with it. Much of the discussion in this chapter has focused on national and international indicators, but to evaluate commercial and community performance it is often necessary to work at more localised spatial scales. Making sense of many terrestrial environmental indicators also requires longer time trends than those needed for economic activities. Social time trends need to have a generational basis (a minimum of 25 to 30 years), but to track changes in climate, or soil properties may require 200–500 years. Interpreting trends in a suite of indicators that are drawn from social, economic and biophysical realms therefore requires some type of sliding temporal scale, and the capacity to scale up and down spatially from local to national.

Spatial scaling is now routinely handed over to technical solutions that involve geographic information systems (GIS), management of very large data sets, coupled with analytical devices such as decision-rule expert systems. Temporal scaling requires statistical interpolation tools combined with process-based modelling and scenario building for future options analysis. The operation of such complex interactive systems then becomes the preserve of specialist units, both in government and the corporate sector. They are costly to run, data hungry, and their results are difficult to challenge by outsiders who do not have the same access to the data sources or resources to re-run the analyses.

AUSTRALIAN INDICATOR SELECTION FOR MANAGEMENT APPLICATIONS

The 1996 Australian Academy of Science Fenner Conference on the Environment provided a convenient moment to check how Australian indicator development was progressing. Entitled *Tracking Progress* (Harding 1996), it was concerned with assessing how far indicators and accounting systems were able to link economic and environmental activities in society, both at a macro (national) scale and at a micro (corporate) scale. It reflects the particular preoccupation of how to develop a value system for environmental assets in a world obsessed with the neoclassical values of economics in which the environment had been largely taken as a free good. This period was marked by the development of indicator system frameworks, which became grouped round the two or three well received models described earlier in this chapter.

Conference proceedings demonstrated that considerable work had been done not only in government, but also by parts of the private sector that were being required to report on their environmental and social programs as well as on their commercial profits and losses. The development of performance indicators that cover such subjects as occupational health and safety standards, environmental and quality assurance audits in both government agencies and the corporate sector may be viewed as an early form of 'triple bottom line' accounting. It is still a somewhat cursory account, however, in most company reports. Strengthening of this type of accounting came as the result of changes to federal law regulating the activities of companies and corporations in the early 1990s.

In comparison with the somewhat contentious gains made in the development and use of national and international sustainability indicators, real progress has been made over the past decade in integrating environmental and quality assurance systems as management tools in the primary and manufacturing industries in many developed countries. These activities have been carried out at localised scale, either within individual companies, local and district governments, or across a particular sector. Their adoption has been widespread in the manufacturing, mining and energy sectors in Australia, with rather more patchy uptake in agriculture, forestry and fisheries.

Many of these are having a positive impact on their immediate environments. This move has not, however, been in response to policy-led activities such as Agenda 21 or the development of sustainability monitoring. Rather it has been part of a worldwide process, driven by increasingly sophisticated consumer patterns in Western societies, and the somewhat chequered progress of trade liberalisation and international price competition. Thus, primary and manufacturing sectors, although a smaller proportion of the total Australian economy, dropping from 25 to 16 per cent of GDP between 1989 and 1999, have literally cleaned up their act, to provide products that can be authenticated as high quality and produced in clean, safe environments.

The initial impetus for such schemes, characterised by the International Standards Organisation (ISO) 9000 series, focused on the factory and processing plant, to ensure a guaranteed product quality. The more recent ISO 14000 series extends scrutiny backward to the environmental conditions of the provision of raw materials and forward to the final disposal of wastes. While the ISO series are costly, comprehensive and not easy for small businesses to implement, a plethora of alternative certification schemes has sprung up in many industries to provide local branding and accreditation schemes that are considered to give either a market niche, access or price premium. Adoption of best management practice (BMP) in Australian rural industries has, for example, been greatest in those industries that are competing for premium export markets; they include the bottled wine portion of the wine industry, processed dairy products, export-quality fresh fruit and vegetables, and certain red meat export lines.

Increased community concern for environmental standards and management in areas that may affect human health and safety has also provided stronger political endorsement of more stringent and wide-ranging regulatory policies across most Western countries. Environment Protection Authorities have been established in all Australians state and territory jurisdictions during the 1990s, with greater powers to inspect, monitor and regulate activities that affect water, air and food quality.

These regulatory powers, however, are patchy in their impact on different industries. Substantial controls exist in the manufacturing, registration and labelling of agricultural chemicals for example; conversely there are no general requirements on manufacturers, distributors or farmers to report where, what and how much of these chemicals are actually used in different environments. Manufacturing and processing industrial plants are highly regulated in relation to their emissions, waste streams, and safety standards, but widely dispersed agricultural practices that can lead to wind and water erosion, destruction of native vegetation and habitats, water pollution from nitrates and phosphates, are much more difficult to monitor or regulate, both practically or politically.

In 1997 the Australian National Pollutants Inventory was established to provide a comprehensive register of those industry activities that are the sources of larger volumes of emissions. Eventually it will require some 80 categories of industry to provide logbook type assessments of emissions to water, air and land for several hundreds of chemical compounds, using standard methods of calculation. The type of database that will develop from the National Pollutant Inventory, openly accessible on the Internet, following a uniform set of reporting and calculation methods, is of enormous value to establishing a quantitative, factual set of indicators. These can be used for many different purposes, just as economic indicators of financial performance are used — for trend analysis, comparisons between sectors, regions and jurisdictions, for identifying critical thresholds and threatened locations, planning future social needs relative to resource use, and in support of trading standards.

HOW FAR SHOULD INDICATORS BE AGGREGATED?

While progress can seem slow in an area so complex as sustainability, there has been a significant increase in the acceptance of the validity of an ESD value-set within many governments over the past decade. To my mind, this is really the triumph of the use of indicators, not their use in policy development, nor their application to management, useful though these may be.

The most valuable contribution has been to draw attention to values other than the obsession with the economic bottom line, and the simplistic assumptions that bedevilled the neo-classical monetarist policies of the 1980s. Some of the continued interest and activity in pursuing the concept of sustainability has undoubtedly come from the greater appreciation of the increasing threat to the environment that overzealous economic activity brings with it. Consumer power and voter power have become equally known and accepted reactions to the individual's feeling of helplessness in the face of giant corporations or unscrupulous politicians. This is why the sustainability indicator systems that involve a wide range of stakeholders are more likely to have lasting influence than those that stem from government alone. The increasing use of a much wider range of statistics, indicators and their trends has also put much more information on a multiplicity of topics into the minds of whole communities.

The tyranny of the unwise or meaningless indicator, however, is also one that threatens to submerge the value of more thoughtful and useful work. This is largely exemplified by the tyranny of the mean, where an average value (such as one numerical value for water quality for a continent) is so meaningless that its use should be condemned. It is a pity that so much current work is devoted to national endeavours and territorial boundaries. Is it sensible to consider one fifth of the world's population alongside tiny populations of less than a million, in the same suite of comparisons? It will be interesting to see whether the rapid increase in information sources based on natural environments, which use topographic boundaries, atmospheric systems, or marine zones, starts to push conventional administrative envelopes. These developments take advantage of the enormous power offered by advanced computing and geographic information systems to report along different regionalisations that have more relevance, at least to environmental systems.

NOTES

1   Published by WRI in late 2000 as Pilot Analysis of global ecosystems, agroecosystems, forest ecosystems, coastal ecosystems, grassland ecosystems and freshwater ecosystems.

2   The HDI's indicator of 'educational attainment' is measured as adult literacy, weighted as two thirds of the value, and combined primary, secondary and tertiary enrolments, weighted as one third value.

3   'Triple bottom line' accounting refers to the use of monetary equivalents for measuring the value of natural and human capital, as well as economically generated capital.

MODELLING PHYSICAL REALITIES:DESIGNING AND TESTING FUTURE OPTIONS TO 2050 AND BEYOND

BARNEY FORAN AND FRANZI POLDY

INTRODUCTION

PUBLIC POLICY, WORLD VIEWS AND ANALYTICAL APPROACHES

Public policy analysis and decision making implies choice: the future is not pre-determined but can be influenced by what we decide to do. There are many alternatives from which we might choose and the choice to do nothing, is a wilful one (Robbert Associates 2000). Within this context, some of the more difficult issues of public policy involve balancing longer-term societal interests and shorter-term individual or private interests. This is particularly so in the case of public policy concerned with the environment, natural resource management and public health. Many of these problems involve externalities — situations where the activities undertaken by one individual or group in pursuit of its objectives, have adverse unintended consequences for other individuals, groups or society at large. Some of them are characterised as problems of the commons where the lack of a clear and just system of property rights, decouples the linkage between shorter-term opportunities, from a longer-term possibility of a run down in system function or productivity. The issues of global climate change and marine fisheries are typical examples where shorter-term expediency leads society to overload the waste assimilation capacity of the atmosphere, or over-harvest particular fish stocks in wild fisheries.

Scientific disciplines assume that they should contribute a key component to both the identification and the resolution of environmental and resource management issues. Where common resources are at stake, science is able to quantify the state of the common resource, as well as possible pathways along which those resources might evolve. Science might estimate concepts such as sustainable yield as well as providing the basis for technologies that might reduce externalities, or increase productivity. However science as a discipline sometimes finds it difficult to properly inform policy analysis, since this requires a fuller understanding of how institutional and political systems might change in response to a particular policy innovation. Various processes such as expert panels, computer modelling and community workshops have been used to help bridge the gap between science and policy development with different levels of success. Systems simulators are a further development in the approaches that science has to offer. They combine observations of the past states of the system (the history) with a scientific understanding of the processes that drive the system. Together, the history and the process understanding can provide the foundation for active learning on how the simulated system responds to policy intervention and innovation.

The success of system simulators in guiding integrated policy advice over the last 40 years has been mixed. The scenarios developed and tested by the Club of Rome 'World' models in the early 1970s (Meadows et al. 1992) were widely interpreted as predictions and were later judged to be incorrect, particularly in regard to resource depletion issues. However many large systems simulators of global climate systems that link back to population growth, energy use, agriculture, forestry and water use have gained wide acceptance in global science and policy circles. A new era appears to be emerging where policy deliberations are again open to approaches of this type.

Figure 8.1
An organising framework of four worldviews within scientific disciplines that determine how the realities within the physical economy are understood, analysed and acted upon.

Source Foran and Poldy (2000)

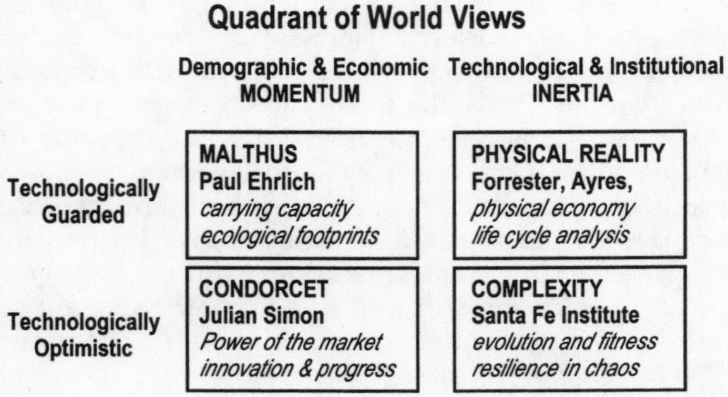

Quadrant of World Views

	Demographic & Economic MOMENTUM	Technological & Institutional INERTIA
Technologically Guarded	MALTHUS Paul Ehrlich *carrying capacity ecological footprints*	PHYSICAL REALITY Forrester, Ayres, *physical economy life cycle analysis*
Technologically Optimistic	CONDORCET Julian Simon *Power of the market innovation & progress*	COMPLEXITY Santa Fe Institute *evolution and fitness resilience in chaos*

However this fledgling new era of policy analysis still has strong linkages back to core debates that were being conducted nearly two centuries ago. In an effort to tease apart some of the foundations of current and future policy debating platforms, four quadrants of worldview, and analytical paradigm are proposed (Figure 8.1). The four quadrants are derived from analytical views that are either technologically guarded or technologically optimistic, and whether those world views are guided by an understanding of the momentum embodied in population growth and economic growth, or the inertia embodied in national infrastructure and societal institutions.

Being guarded or optimistic about the prospects of technological innovation and advancement does not constitute a right or a wrong. Rather it classifies the source of the data used in supporting these world views, and the types of analytical procedures uses to promulgate them. The understanding of momentum (quantity of motion) in an economic and demographic sense is based on an understanding of the structure of human populations and monetary economies, the potential for growth therein, and the time periods required before a different structure of population or economy can be attained. The understanding of inertia (sluggishness) in an infrastructural and institutional sense is derived from the observation that infrastructure (houses, roads, bridges, power plants) and institutions (courts, laws, parliaments, schools, business affiliations) have a wide range of characteristics that enforce their current structure and limit the rate of change. This inertia restricts the capacity of new technologies and new modes of organisation to replace the current status quo. The mapping of a particular policy approach or method of analysis into a system such as the four quadrants helps characterise the methods used, the disciplinary base of the analysts and ways in which the results will be extended into policy relevant discussions.

When the Reverend Thomas R. Malthus first wrote his essay, *A Summary View of the Principle of Population* as a supplement to the 1824 version of the *Encyclopaedia Britannica* (Mentor Books 1960), he would not have predicted that the debate would still be raging at the start of the second millennium. Scientists such as Paul Erhlich (1968) and Lester Brown (Brown et al. 1998) still propose that continuing population growth and linked lifestyle and resource consumption pose a serious threat to the ecological integrity of world ecosystems. The analytical methods used in this quadrant (technologically guarded, demographic and economic momentum) are well versed in demography, ecology, pollution generation and production of natural resources. The concepts of human carrying capacity (Cohen 1995) and ecological footprints (Wackernagel and Rees 1996) are good examples of concepts developed in this quadrant. These analyses are usually well based in data but are relatively simple and often attract criticism because they are static, they do not give a way forward in policy terms and they ignore much of humankind's history of innovation and progress.

In the same era that Malthus was promoting the world views of the first quadrant, the views of the technologically optimistic group in the demographic and economic momentum column were being promoted by the Marquis de Condorcet (University of Berkeley 2000). He espoused that the richness of the human spirit had the potential to overcome all odds, and that there was no limit to humankind's capacity to invent and solve. The seeming reality of these views are repeated continually today and many of Condorcet's disciples, most notably Julian Simon (1990), have won a number of critical debating points over the their technologically guarded colleagues. The analytical methods of this group generally include the many approaches used in economics, most notably the computable generalised equilibrium models at the heart of national decision making on macro-economic areas in most developed economies. In Australia these include the MONASH model developed at Monash University, the TRYM model used by Commonwealth Treasury, the Murphy Model used by a private firm Econtech and the Salter Model used by the Federal Department of Foreign Affairs and Trade (EPAC 1994). Some criticisms directed at this group include the absence of equilibrium in most functional economic and natural systems and that the behavioural assumptions around the concept of elasticity do not have much validity beyond the short run.

The third quadrant, characterised as technologically guarded and attuned to the inertia in most infrastructure and institutional systems, is typified by the work of Ayres (1998), Slesser et al. (1997) and Forrester (1961). Their analytical approach has two key components not found in the first two quadrants. The first is that the economic and social worlds lie within the physical world and must therefore eventually conform to physical laws. These laws include the laws of thermodynamics and mass balance, which eventually will impose constraints on the optimism of the proponents of the world views found in the second quadrant. The second component is the use of dynamic systems modelling techniques to avoid assumptions that most human systems seek some form of notional equilibrium where all forces are in balance. These modelling approaches also ensure that important forces such as population growth and economic growth are linked to the biophysical realities of resource requirements and the production of waste and pollution. Critics of these approaches question the degree to which changes or improvements in human behaviour, substitutions between materials and the effect of continual innovation are excluded from modelling considerations. The modelling approaches discussed in this chapter lie mainly within this third quadrant.

The fourth quadrant includes those that are both technologically optimistic and aware of inertias in infrastructure and institutions. They are typified by the complex systems research currently under way in the Santa Fe Institute in the United States (Santa Fe Institute 2000) and made popular by books such as *Complexity* (Waldrop 1994). The complexity quadrant brings together seemingly disparate groups such as evolutionists, economists, ecologists and pure mathematicians to help impose order and understanding on complexity and chaos from the level of genes to money markets to climate systems and the intricacies of the future human mind. The overall methodology employed by this quadrant is hard to typify beyond being based on deep mathematics, agent based modelling and a number of other approaches. A current criticism of the approach might be that it is difficult to understand and apply outside its immediate research environment.

POPULATION-DEVELOPMENT-ENVIRONMENT STUDIES IN AUSTRALIA

Modelling for environmental sustainability was stimulated in Australia by the continuing population debate. The concepts of population targets and carrying capacity have a long history in Australia starting in the 1920s when a Sydney University geographer Thomas Griffith Taylor set Australia's estimated carrying capacity at 65 million people and later reduced this estimate to 20 million people (Cocks 1996). During the 1980s and 1990s there have been several national inquiries on population, the most recent of which was the Jones Inquiry (Long Term Strategies Committee 1994), which stopped short of recommending a national population policy (Cocks 1996). By default, Australia's population seems to be moving towards a more or less stable population of around 23–25 million people in one to two human generations' time. During the 1990s, the national population debate evolved to include a wide range of issues such as resilience of ecological systems, material consumption levels, sustainability issues and population size as a determinant of domestic market efficiencies and Australia's place in world affairs.

It was against this background that CSIRO, a national science agency, initiated a strategic project to underpin the population debate with scientific analysis. The project's initial aim was to focus on the environmental aspects of population impact with particular emphasis on the quality and quantity aspects of water, soils, biodiversity, atmosphere and natural amenity. Initially the work proceeded on a traditional scientific route where plans were made to examine the effect of population on water resources, land resources and so on. However because of the complex linkages between all sectors of society and the economy, the traditional approach of defining tight boundaries around a well defined problem prior to analysis was judged difficult to implement. In addition, the project was challenged with a future orientated and long-term topic that required integrated advice and a range of possible solutions. At this time the project became aware of two important methodologies. The first was the work of Godet (1991) and his work on 'strategic prospectives' and thence the use of fore-sighting and scenario development by multinational companies such as Royal Dutch Shell. The second was the implementation of population-development-environment simulators, particularly the work by IIASA in Mauritius (Lutz 1994), the physical analysis paradigm using the design approach (Gault et al. 1997) and the embodied energy approach of Slesser (1992, 1997) and colleagues.

The project design then evolved under the strategic aim of 'influencing national policy agenda in regard to population policy and the impact of humankind on the environment'. Two linked themes of work emerged. The first was scenario development where the project aimed to develop a number of robust and well-documented national scenarios that could lead and inform the debate on national development and sustainability. Three scenarios, Economic Growth, Conservative Development and Post Materialism have been now released in book form (Cocks 1999). The second theme of work developed in order to underpin this broader more qualitative debate with quantitative analyses. Within this theme, two system simulators were developed based on different approaches of physical analysis. One of these OzEcco (Foran and Crane 1998), used the embodied energy approach of Slesser (1992, 1997) to construct a top-down and aggregated simulator of Australia's physical economy. This analytical approach assumed that the delivery of goods and services to a domestic economy, and the human population therein, is a function of the extraction, delivery and efficiency of use of energy resources, most of which are fossil fuelled based.

The second simulator, the Australian Stocks and Flows Framework (ASFF), was a disaggregated set of linked models that accesses a database describing the last 50 years of Australia's physical function or physical metabolism. The design approach used in ASFF described below is philosophically attractive for two reasons. Firstly, it treats all aspects of physical function as separate entities (crops, animals, people, cars, steel production, chemical production) and allows a detailed treatment of vintaging or age for most big-ticket items of physical infrastructure. Secondly, the physical functioning is retained within the modelling code. The management and policy decisions that guide this physical functioning are retained as part of a scenario under development and testing by the user or policy analyst. Gault et al. (1987) describe the design approach thus:

> The design approach is a philosophy for building computer based simulation frameworks, which represent socio-economic systems, and for using the simulation framework to design alternative futures through repeated simulation. It is the exploration of alternative futures by the

user, who forms part of the system, which distinguishes this approach from that of macro-economics with its emphasis on prediction. The exploration and the involvement of the user result from the absence of optimization or equilibrating mechanisms in the physical representation of the socio-economic system. This ensures that the user, working alone or with the aid of a model of decision processes, controls the system. The policy decisions necessary to exercising this control are required to be explicitly stated, and they form a record of how the future, resulting from the simulation, was arrived at.

A brief description will be given of the IMAGE global change model as an example of more global approaches to the physical modelling approach (Alcamo et al. 1994, Alcamo et al. 1998). At a lower scale, the chapter will then describe both the OzEcco embodied energy model and the Australian Stocks and Flows Framework (ASFF) as examples of modelling approaches designed specifically to deal with the long-term challenges of sustainability at a national level. It will give some current examples of model use within policy and science process-es. It will also note the challenges for these analytical approaches in achieving a goal of 'influencing national policy'. The chapter will end with some partial insights into the many conundrums that face inte-grative modelling of the physical economy approaches of this type.

MODEL DESCRIPTIONS

MODELS OF GLOBAL CLIMATE CHANGE

The issue of global climate change has stimulated the development and use of a large variety of modelling frameworks, some of which deal comprehensively with one issue such as carbon metabolism at a global level, and others that attempt to integrate all important issues in an approach termed 'integrative assessment' (Goudriaan et al. 1999). One such integrative assessment model is IMAGE (Alcamo et al. 1994, 1998) developed by the National Institute for Public Health and the Environment in The Netherlands. It combines three distinct areas of the Energy-Industry system, the Terrestrial-Environment system and the Atmosphere-Ocean system (Figure 8.2). The first two of these sys-tems are central to the nationally scaled models described next in this chapter, but they lack the Ocean-Atmosphere system. The key differ-ence between this approach and the latter ones is that of scale. In the IMAGE approach the physical metabolism of the entire globe is mod-elled in 13 different regions whereas the national models deal with one nation that is modelled in many sub-divisions.

The IMAGE model is used to link the effects of human manage-ment through the full chain of physical processes that run the globe. Thus increasing population and affluence causes land use change and increasing energy use, all of which increase the emissions of carbon dioxide and other greenhouse gasses such as methane. These emissions cause changes in function of the Earth-Ocean system leading to changes in rainfall and temperature that, over the duration of the model simulations can feed back to affect sectors such as agricultural productivity and water yield from catchments. While models such as IMAGE are always run in scenario testing mode, their modes of usage include both prediction as well as back casting (or hind casting).

In prediction mode, assumptions are made that cover the full range of possibilities for the driving forces and the simulation outputs include issues such as atmospheric concentrations of greenhouse gases, temperature changes, rises in sea levels and changes in agricultural productivity. Because the future is indeterminate, such assumptions usually include a full range of sensitivity testing so that ranges of error or probabilities of outcomes can be measured. The prediction mode can lead to the next stage of model simulation, that of back-casting. In back-casting, important assumptions that drive global climate change (population growth, fossil energy use, land use change) are altered in an attempt to find a combination of feasible settings that reduce or change the nature or the severity of the initially simulated outcome. Using both the prediction and back casting modes in tandem can lead to sets of linked insights or understandings that can lead to new and improved

Figure 8.2
A schematic representation of the IMAGE 2.0 integrated assessment model

Source Alcamo et al. (1994)

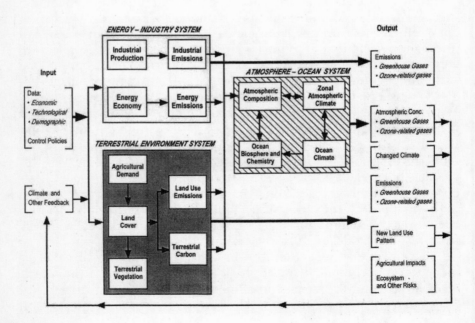

policy design and implementation. At some stage in the global simulation process, these broader insights must be applied to a national scale, where deeper and more insightful analyses are required to accommodate the social and political changes required. At this stage the next level of modelling is required.

THE OZECCO EMBODIED ENERGY MODEL

The OzEcco model is designed to integrate the driving forces of population, lifestyle, organisation and technology and explore their possible impacts on the environment within the context of Australia's physical and economic structure. It is a systems dynamics representation of Australia's national function based on the philosophy of embodied energy analysis. The structure of the national economy and the energy accounts have been integrated so that capital stocks are expressed in a physical measure of petajoules of embodied energy rather than constant dollars. The activities within the economy have been expressed as energy flows, again in petajoules. In this way economic activity has been converted to physical activity, which is consistent with the first and second law of thermodynamics. All economic transactions are represented by the physical transformations that underpin them. This representation is consistent with the long-term physical processes that are central to the functioning of any modern economy.

Figure 8.3
A diagram of the central growth-determining loop in the OzEcco model*

SOURCE Foran and Crane (1998)

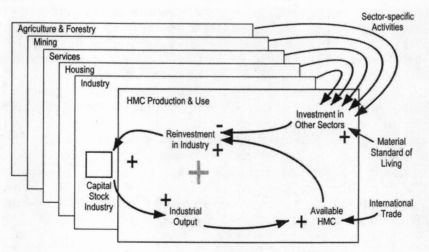

* The aggregated industrial sector depicted here as the core resource on which growth depends. The processes of fixed- or human-made capital (HMC in diagram) are depicted as an influence diagram, illustrating the main causative features represented in the model. The total human made capital available is the sum of imports and domestic production.

Conceptually the model has five broad components: natural resource stocks, the transformation sectors, consumption activities, pollution generation and whole system indicators. The core modelling concept is that access to, and transformation of energy (typically stocks of fossil fuel) are the determinants of physical growth in a modern industrial economy. Thus all goods and services are seen in terms of their embodied energy content. Some sectors such as domestic housing act as long-term accumulators of fixed energy capital (expressed in embodied energy terms), whereas personal consumption dissipates embodied energy quickly. The concept is shown in Figure 8.3. The capital stock of industry (expressed as embodied energy) is a primary focus that creates human-made capital through its contribution to other sectors such as agriculture (fertiliser, machines), domestic housing (bricks, carpets, stoves) and so on.

The rate at which industry can grow in any one year is limited by the contribution that this sector makes to other sectors of the physical economy and the consumption activities of the population at large. These are both negative feedbacks that act as brakes on the rate at which the physical economy can grow. The effects of international finance can be both positive and negative. Exports are negative in that they reduce the amount of physical capital (embodied energy) that can be applied nationally. Physical imports and monetary inflows are positive influences in that they increase a nation's ability to perform physical work. All of these factors are linked in a systems dynamics framework. The system is set to grow as fast as is physically feasible (governed by the first and second laws of thermodynamics) in a physical economy that is constrained by the availability of fossil energy, the requirement to maintain national infrastructure and personal consumption activities. Global monetary flows (for example, balance of payments and international debt issues) can be interpreted as flows of virtual energy that might override some resource and infrastructure issues in the physical economy for the short term.

For a number of reasons, acceptance of the OzEcco approach by both the science and the policy community has not been assured. The use of a numeriare such as embodied energy is difficult for some policy analysts to accept. However two recent developments in the energy and greenhouse area of the physical economy have increased the potential use of this type of modelling framework. The first is the acceptance of static analyses of energy embodiment using input-output tables of the monetary economy to determine the energy use and greenhouse gas generation by different sectors of the economy (Lenzen 1998). The second is the use of OzEcco in designing transitions towards a biomass based transportation cycle that is attracting a degree of national policy interest (Foran and Mardon 1999).

THE AUSTRALIAN STOCKS AND FLOWS FRAMEWORK (ASFF)

General description

ASFF is a highly disaggregated simulation framework that keeps track of all physically significant stocks and flows in the Australian socio-economic system. In this context, stocks include people, livestock, trees, buildings, vehicles, capital machinery, infrastructure, land, air, water, energy and mineral resources. They are disaggregated, as appropriate, according to their physical characteristics and importantly, age or vintage. Flows, resulting from physical processes of many kinds, represent the rates of change of stocks and constitute the development of the system in more or less desirable directions.

The framework consists of a simulation model and a database. The simulation model consists of 32 hierarchically connected modules or calculators that cover the accounting and physical processes of demography, consumption, buildings, transport, construction, manufacturing,

Figure 8.4
Hierarchy of calculators in the Australian Stocks and Flows Framework. See Figure 8.5 for information flow between calculators.

SOURCE Poldy et al. (2000)

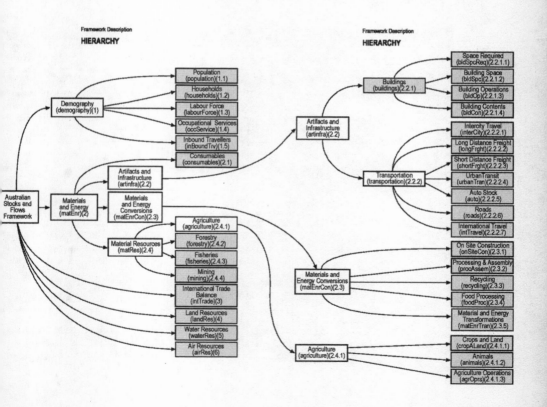

energy supply, agriculture, forestry, fishing, mining, land, water and air resources and international trade. Each calculator deals with the stocks and flows relevant to a sector and with the physical processes through which they interact.

Calculator assumptions are based on technical and scientific understanding of the processes involved, and are intended to provide a plausible representation in physical terms of the workings of the sector. Indeed, it is a criterion of validity for the calculator, that a professionally informed person should be able to follow the structure of the representation and conclude that it and the values of parameters are plausible and appropriate to the level of aggregation of the treatment.

An overview of the whole framework is given in Figure 8.4, where the arrows link calculators arranged in functionally similar and hierarchically related groups (note that the arrows do not represent sector linkages or information flows — these are shown in Figure 8.5).

Figure 8.5.
One way information flow (vertical arrows) between calculators (horizontal lines) of the Australian Stocks and Flows Framework. Shaded calculators receive only exogenous input (no arrowheads on shaded lines).

SOURCE Poldy et al. (2000)

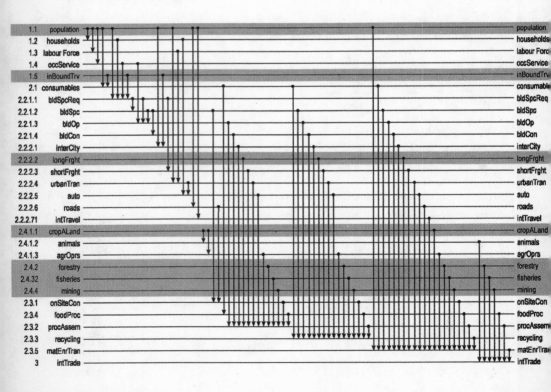

The model calculators

In Figure 8.4, the clear boxes with bold borders represent hierarchical groupings, and shaded boxes represent calculators. At the highest level, the Australian socio-economic system is conceived of in terms of people (Demography) and the physical needs of their way of life (Materials and Energy). Population is an important driver in the framework and, other things being equal, more people require more materials and energy. Other things are not necessarily equal, and one of the goals of ASFF is to explore the interplay and trade-offs among Population, Lifestyle, Organisation and Technology — the so called PLOT factors.

The five Demography calculators deal with population (including overseas and internal migration) and issues that depend directly on population and its distribution over age, sex and location: education needs, morbidity and health needs, internal travel, household formation, labour force participation, demand for personal services and inbound tourism. Population and inbound tourist numbers are independent drivers in the framework, that is, the parameters that determine their level and growth are specified exogenously. Information from these demography calculators is passed to later calculators and used to determine the requirements for infrastructure, goods and services of all types.

The Consumables calculator determines the need for food and other consumable items directly from population (including overseas visitors) on a per capita basis. The four Buildings calculators use information from demography to determine the needs of the population for residential, commercial, educational, health care and institutional buildings. Seven calculators deal with various aspects of Transportation. Broadly these cover domestic passenger and freight transport in urban and rural areas. Separate calculators deal with the car fleet, roads and their maintenance, and fuel for international travel. In most cases, a transport task is determined in relation to demographic parameters and, with the help of load factors and average yearly distance travelled, the task is translated into a need for vehicles. The Material Resources calculators describe production processes in the primary industries: agriculture, forestry, fisheries and mining. Like population, tourism and long distance freight, they are independent drivers in the framework and receive no information from earlier calculators. Their planned levels of production are specified exogenously because most of the produce of Australian primary industries is destined for export.

Agriculture is covered by three calculators that deal with crops and land, livestock and agricultural operations in each statistical division. Cropping deals with the areas of land devoted to each of ten different crops (or land may remain fallow or idle), the impact of cropping activity on four indicators of soil quality (acidity, dryland salinity, irrigation salinity and soil structure) and the effect on yield of genetic improvements to crop varieties, the application of fertiliser and irrigation and of declining soil quality due to the cumulative effects of previous cropping. The Animals calculator deals, in each statistical division, with the stocks of animals of different types, the quantities of animal products they yield and their feed requirements in terms of crops and area of grazing land.

Forestry deals with 15 different types of forest managed under regimes that vary from full protection to clear cutting and managed plantations. Fire frequency and tree growth and survival rates are taken into account. Inventories are kept of land areas and of tree numbers and wood volumes by age. Fisheries deals with both wild fishing and fish farming. Wild fish stocks vary in response to their own natural rates of reproduction and mortality and to the level of fishing. Each fishery can sustain some moderate level of fishing but, if overfished, the stock collapses to levels at which catch per unit effort no longer warrants fishing. Fishing effort is allocated among fisheries in an attempt to meet planned production levels at minimum effort.

Mining covers exploration for mineral and energy resources, evaluation and classification of resources as reserves, and extraction of minerals and energy materials to meet planned production. Resources found is the current estimate of the nation's total endowment of a material. Unless augmented by new discoveries, cumulative production will never exceed this quantity. The Materials and Energy Conversions group of calculators covers construction, manufacturing and energy supply. Its calculators deal with the need for materials, energy, goods and infrastructure identified in earlier calculators. Processing and Assembly consolidates the requirements for vehicles, machinery, building contents and operating goods of all types from previous calculators and, allowing for imports and exports, determines the level of domestic production of these goods. Recycling consolidates all discarded goods, vehicles and machinery and determines the proportions to be recycled or disposed of to landfill. The material content of the recycled fraction is determined from a knowledge of the material composition and vintage of the goods and vehicles. Material and Energy Transformations ensures that the needs of the whole economy for materials and energy are met.

The International Trade calculator consolidates domestic production and domestic requirements for primary materials, secondary materials, vehicles and machinery, intermediate and final demand goods and determines import and export quantities. These are combined with a set of import and export prices and an interest rate, to determine the value of the trade flows, the current merchandise trade balance in nominal dollar terms and its contribution to the international debt (or surplus) again in nominal dollar terms. Finally, Land Resources, Water Resources and Air Resources consolidate information from the whole framework into accounts that provide an overview of the state of these important resources.

The framework is grounded in a database for the historical period (the 50 years to 1991), which is complete (all data gaps are filled), and where variables are consistent with each other and with the assumptions in the calculators. These assumptions are based on technical and scientific understanding of all the processes required to describe physical stocks and flows underneath the Australian socio-economic system. At the most basic level this ensures that fundamental requirements such as the conservation of matter and energy and the laws of thermodynamics are observed. For particular calculators, the assumptions are required to be consistent with a specialist's understanding of the processes involved.

Calculator linkage, feedback and tensions

The calculation linkages are shown in Figure 8.5 where arrows flow downwards only indicating that feedbacks caused by demand and supply imbalances are controlled by the user, who separates control space from design space. In order to calculate the quantities demanded within the physical economy, the population calculator (1.1 in Figure 8.5) passes down:

- the requirements for households (1.2) through an age and sex determined household formation rate;

- the availability of a labour force (1.3) through an age and sex determined participation rate;

- the demand for employment in non-physical sectors of the economy, such as services (1.4) as a proportion of the total population;

- consumables such as food, plastics, paper, pharmaceuticals and chemicals (2.1) on a per capita per year basis;

- the demands for building space (2.2.1.1), intercity travel (2.2.2.1), urban transit (2.2.2.4), roads (2.2.2.6), international travel (2.2.2.7) and material transformations (2.3.5).

This process is continued down the hierarchy of calculation procedures given with a complete set of quantities demanded by the population driver and the subsequent flow-on effects. In order to supply the quantities demanded, production or control variables are set in the primary material sectors (agriculture, forestry, fishing, mining) or the international trade sector, so that the quantities demanded by the population might equal the quantities supplied over the period of the simulation.

Figure 8.6
Content and information flow between control space and information space in reality and in the Australian Stocks and Flows Framework

SOURCE Poldy et al. (2000)

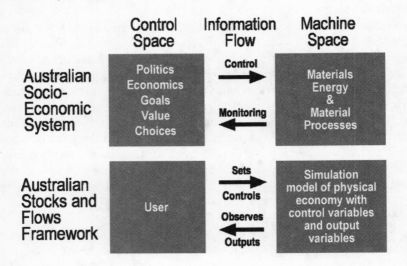

The design (and testing) approach that lies behind the implementation of the ASFF model distinguishes *control space* from *machine space* (Figure 8.6). Control space is occupied by the user or analyst who makes assumptions on the basis of current knowledge and future expectations and then alters control variables in the ASFF model. Machine space is occupied by the modelling code and the equations that describe the processes that drive the physical economy. This is the domain of materials, energy and physical processes. What happens in machine space depends on physical laws, but it also depends on choices made in control space according to people's values. However, people's control of the physical world is imperfect both because the physical world is very complex and because their goals and values conflict with other people's. From control space, the analyst can monitor what goes on in machine space during model simulation and evaluate the outcomes according to goals and values set by a research group or a client. In practice the iterative nature of design and testing can be slow and spasmodic as simulation outcomes are delivered to clients as documents with scenario graphs and written interpretations. In theory a policy client and a model analyst could sit together at the computer screen and increase the speed of design and learning.

In the design approach of the ASFF, only the physical processes in machine space are modelled. The user occupies control space, observes the situation in machine space and makes decisions about the settings of the control variables. The user is therefore an integral part of the feedback loop, acting as a proxy for society and its political and economic agents, and is in a position to learn a great deal about the system behaviour.

Resolving tensions (imbalances between quantities demanded and quantities supplied) may be obligatory or optional. The difference is that if a tension indicates a physical or accounting inconsistency, it must be resolved. For example, if insufficient primary energy is supplied to meet electricity and transport requirements, then its supply and delivery must be increased. Another form of tension might indicate the failure to meet some non-physical goal or desirable criterion. In this case, its resolution is judged to be optional as illustrated by an imbalance between the labour demanded and the labour supplied. If there is more labour supplied than is demanded, then this is called unemployment and the scenario is still physically feasible. If there is more labour demanded than supplied, then the production goals might be regarded as infeasible. Production goals might have to be decreased, or the labour force increased.

APPLICATIONS AND RESULTS

THE OZECCO EMBODIED ENERGY MODEL

For a policy client interested in alternative landuse scenarios that might help re-mediate landscapes suffering from dryland salinity, scenarios were designed within OzEcco to implement the production of alcohol fuels from woody biomass (Foran and Mardon 1999). There were a number of assumptions that underpinned this methanol production scenario shown as follows:

- The scenario would aim to supply 90 per cent of Australia's total oil requirements specifically to meet 100 per cent of the requirements for transportation fuels.

- The feedstock share would be 100 per cent woody material from plantation biomass resources that are currently managed as forests with a 20-year rotation and an average 20 m3 per year mean annual increment.

- Approximately 60 per cent of the woody biomass would be derived as logs and the remainder as branches and waste wood.

- The rate of plantation biomass establishment (basically forests) would be 400 000 hectares per annum.

- The capital cost in constant dollar terms of the methanol plant was $50 million per petajoule of production capacity and the lifetime of plant was 20 years.

The top-level indicators produced by running the OzEcco model with these scenario assumptions are shown in Figure 8.7. The simulated growth rate in GDP for this scenario tracks with or above the base case for the duration of the simulation. The first drop due to oil depletion is avoided and the second drop due to gas depletion is not as large. The per capita affluence measure (gigajoules of embodied energy per capita per year) tracks with the base case until 2030 and then takes a higher trajectory. The energy intensity of GDP (megajoules of fossil energy per constant dollar of GDP) is decreased by about 30 per cent (from 8MJ per dollar to 5MJ per dollar) by 2050. The emissions of carbon dioxide from the energy sector diverge from the base case after 2005 and rise gradually to 1000 million tonnes per annum by 2050, a reduction of 200 million tonnes per year compared to the base case.

Figure 8.7
Report card #1 for the methanol scenario (Meth-0)*

SOURCE Foran and Mardon (1999)

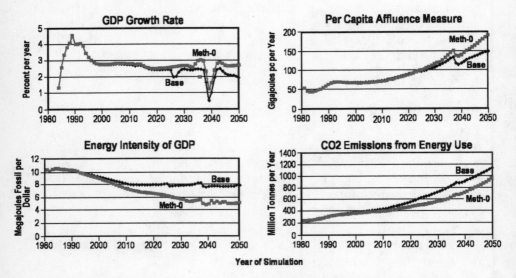

* Comparison made with the base case (Base) showing growth rate in GDP (top left), per capita affluence index (top right), energy intensity of GDP (bottom left) and carbon dioxide emissions from energy use (bottom right)

Analyses such as these are not predictions but test the likely behaviour of the modelled physical economy to policy innovations and new structural designs. A measure of scenario success is the degree to which indicators for a scenario under test diverge from the base case scenario. While the OzEcco model runs on physical processes, it is possible to derive a number of economic indicators such as (nominal) GDP because of the strong relationship in the current structure of the economy between dollar production and fossil energy usage. These relationships are well analysed in Lenzen (1998).

There are literally hundreds of indicators able to be displayed for each scenario run. The current method of displaying these is to form them into a number of report cards for different levels of the physical economy that display four indicators simultaneously. Figure 8.6 displays the macro-level indicators that are then supplemented by more detailed report cards of the operations of the physical sectors that are being restructured in a particular scenario. The decision on what constitutes a successful scenario is a difficult one in a policy or industry context, a problem also faced in Chapter 5 in the choice and subsequent interpretation of environmental indicators. Compared with the indicators commonly used in State of the Environment reporting, the advantage of the physical modelling approach (compared with series of reporting indicators obtained from a wide variety of partially linked national statistics), is that modelling indicators are structurally linked to each other through the operations of the physical economy. Provided that the modelling has a sound philosophical and bio-physical basis, this provides a more thorough basis for interpretation and understanding, as well as a cogent and robust look-ahead capability.

THE AUSTRALIAN STOCKS AND FLOWS FRAMEWORK

The intended influence of the ASFF analytical approach is now shown through three contrasting applications. The first is a single sector approach that concentrates on population issues. The second is a multi-sector approach that links population scenarios to vehicle scenarios and the resultant demand for energy use and generation of emissions. The third seeks to identify possible bottlenecks or constraints to the availability of water in urban situations.

Single sector scenarios

Within this analytical approach, the population calculator is seen as one of the main drivers of demand for of food, paper, water and energy and subsequent flow-on effects and impacts. However there are many analytical insights that are important in their own right particularly in Australia where immigration policy is the main policy lever whereby future population stocks or targets will depend on the degree to which immigration is used to offset declining birth rates. Three scenarios of net immigration (zero, 70 000 per year, two-thirds of one per cent of total population per year, i.e. 0.67 per cent per annum) were combined with the expected declining total fertility rates (from 1.78 to 1.65 children per woman), and increasing longevity (one year life extension for each decade of the simulation to 2051). The results are presented for the years 2051 and 2101 (Table 8.1). Australia in 2051 could be home to 20, 25 or 32 million people depending on its choice of net immigration rate. While a zero net immigration is the policy position of several environmental groups, detailed demographic analysis (McDonald and Kippen 1999) shows that this option produces eventual population decline and substantial falls in the size of the labour force. The analysis within the framework is consistent with the more detailed work, and total population has declined from 20 million in 2051 to 17 million in 2101 under the zero net immigration scenario. From 2051, the 70 000 net immigration scenario increases by 0.4 million people by 2101 whereas the 0.67 per cent per annum net immigration has increased by 18 million people to 50 million people and is still growing at 2101.

Table 8.1
Scenarios for Australian population size in millions based on zero, 70 000 and 0.67 per cent per annum net overseas migration, declining fertility rates and increasing longevity.
SOURCE Foran and Poldy (2000)

Year	Zero Net Immigration per Year	70 000 Net Immigration per Year	0.67 % pa Net Immigration per Year
2051	20.6	25.1	32.5
2101	16.7	25.5	50.6

Higher rates of net overseas migration are assumed to make the Australian population younger. The scenarios modelled here assume that future immigration has the same age and gender distribution as the last decade and the results suggest that zero immigration gives a higher proportion of population over 65 years of age (Table 8.2). On a percentage basis, 27 per cent of the population is older than 65 years in 2051 for zero immigration, versus 25 per cent for 70 000 net immigration and 20 per cent for the 0.67 per cent per annum scenario. While the more detailed analyses of McDonald and Kippen (1999) show levels above 80 000 net immigration do not contribute to the retardation of population ageing, the ASFF implementation of the 0.67 per cent per annum scenario is constantly growing and has specific assumptions surrounding the younger age distribution of the immigration.

Table 8.2
The effect of three population scenarios on the percentage of the population over 65 years of age in the year 2051.
SOURCE Foran and Poldy (2000)

Scenario	Zero Net Immigration per Year	70 000 Net Immigration per Year	0.67 % pa Net Immigration per Year
Proportion >65 years of age (percentage)	27.00	25.00	20.00
Number >65 years of age (millions)	5.65	6.32	6.52

Another insight to the data is given if absolute numbers are viewed instead of proportions. Since the smaller proportions are of larger populations, the absolute numbers and therefore the demand for consumption and specialist services will be greater in the larger populations. There are 5.65, 6.32 and 6.52 million people over 65 in 2051 for the zero, 70 000 and 0.67 per cent per annum scenarios respectively. It is possible that social tasks such as aged care, personal security and pensions will be larger in absolute terms with higher net immigration rates if all other policy variables are kept neutral. The effect of population ageing is distributed differently throughout the states of Australia. A number of lesser populated states are characterised by inflows of younger ages and outflows of older ages, which will maintain a relatively younger population if the internal migration dynamics of the last decade are maintained into the future.

Australia is characterised by a high proportion of its people living in the main capital cities of each state or territory. The challenge of maintaining quality of life, efficient infrastructure and economic productivity is a real one (Newman and Kenworthy 1999) and the potential size of Australian cities is an important national policy consideration. The trajectories of population change in cities show different patterns due mainly to the different patterns of internal migration (Table 8.3).

Table 8.3
Simulated population size in millions for capital cities in states and territories in 1998, 2051 and 2101.
SOURCE Foran and Poldy (2000)

	Estimate 1998 ABS	Zero 2051	Base 70 kpa 2051	0.67%pa 2051	Zero 2101	Base 70kpa 2101	0.67%pa 2101
Sydney	3.986	3.946	5.129	6.977	3.194	5.186	10.908
Melbourne	3.371	3.274	4.290	5.982	2.589	4.269	9.350
Brisbane	1.574	1.992	2.503	3.064	1.704	2.633	4.759
Adelaide	1.088	1.133	1.493	2.070	0.912	1.510	3.261
Perth	1.341	1.697	2.163	2.812	1.442	2.270	4.460
Hobart	0.195	0.204	0.270	0.374	0.166	0.275	0.594
Darwin	0.086	0.073	0.096	0.135	0.061	0.100	0.217
Canberra	0.308	0.247	0.325	0.444	0.205	0.327	0.690

Under the lower population scenario, most capital cities reach a peak population in the period 2006 to 2021 and Sydney and Melbourne then start a slow decline in population size. In the medium population scenario, the city populations start to plateau around 2036 and then stabilise around 2051. For the higher population scenario, the populations of all the major cities will continue to grow until 2051 and beyond. In 2051 under these scenarios, Sydney could have 4, 5 or 7 million people by 2051, and Melbourne 3, 4 or 6 million. Brisbane could vary from 2 to 3 million, Perth from 1.7 to 2.8 million and Adelaide from 1.1 to 2 million. These simulated results reflect the interstate migration patterns of the last 10 years. These patterns may change substantially during the next 50 years as the economic and lifestyle drivers of internal migration alter the flows of people between states. In environmental terms these scenarios of population numbers

Figure 8.8
The stepwise progression of computation within the Australian Stocks and Flows Framework whereby scenarios of population change are linked to the personal vehicle calculator and to total energy use and vehicle emissions.

SOURCE Foran and Poldy (2000)

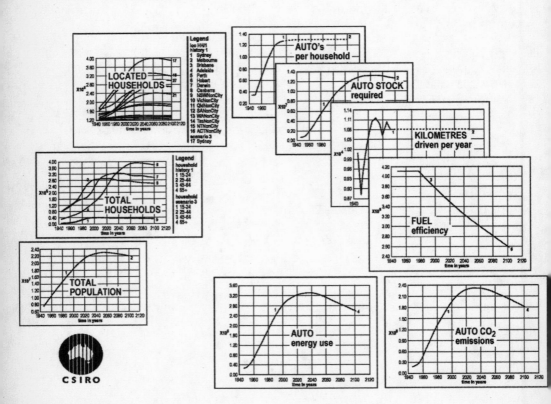

represent the primary pressures in the pressure-state-response frame-work described in Chapter 3. On the basis of these population scenarios, questions for the pressure-state-response framework for future policy contexts could focus on the likely environmental impacts of three population levels with differing proportional representation.

Multi-sector scenarios

While the analyses from any of the ASFF calculators such as demography can be obtained from a wide range of specialist research agencies, it is the onward chain of computation through other parts of the physical economy that allows scenarios to become more technically explicit and useful to policy. For the 60 000 net immigration population scenario, this example shows how population and location parameters are linked to motor vehicle usage, fuel consumption and subsequent vehicle emissions (Figure 8.8).

The driving variable for vehicle ownership in ASFF is the individual household, which locates vehicle ownership in capital cities and regional areas around Australia. Each household is assumed to require 1.3 vehicles, each vehicle is driven 15 000 kilometres per year and the fuel use per kilometre driven declines by 60 per cent over the next 100 years. The analysis from these explicit assumptions shows that the total energy used by the automobile fleet and the subsequent vehicle emissions reach their peak around 2030 and then start to decline. The continued demand for vehicle ownership is built into the continued growth in population and therefore younger households coming into the market for car ownership. At the mature end of the population age distribution, people are living and staying healthy and active longer, and car ownership and usage might be maintained longer than in the past.

There are many ways in which these simulated outcomes might be altered, particularly by technological innovation. Car ownership per household might decline, fewer kilometres per year might be driven, and engine technology might leapfrog the current energy use parameters and solve the problem of vehicle emissions entirely. However Australian lifestyles might dictate that more cars are demanded per household and more kilometres are driven per year. The long time-frames required to alter automobile energy use under these particular assumptions could help frame a policy trade-off where the capacity to overcome the inertia facing technological innovation is judged against the political risks inherent in forcing a change in consumer behavior.

Identifying possible bottlenecks

Australia is a relatively dry continent with high annual variability of rainfall and a reliance on irrigated agriculture for many of its higher value commodity exports such as wine, cotton and dairy products. In the more populated parts of Australia there is competition for the use of water and concerns for both the quality and quantity of future water supplies (Thomas et al. 1999).

In considering the direct requirements for water use by people, significant infrastructure and management issues are associated with maintaining clean catchments and ensuring the chemical and biological quality of water supplies for most major cities. If Australia continues along its present human population track it will have around 25 million people by 2050 and this suggests urban requirements of around 6000 gigalitres of water per year (Figure 8.9). This assumes that water can be transferred from agricultural usage. However if that were not possible for the base case scenario, there is a suffecent range of options in terms of takeback from other uses, industries and water savings that might be instituted in each home to ensure that there is enough water available.

When the other population scenarios portrayed in Figure 8.9 are compared to the base case, the requirements are 2000 gigalitres per year more for the higher scenario with 32 million people and 1000 gigalitres less for the lower scenario with 20 million people. By 2100 however, a number of water availability tensions could appear as the direct population requirements are 12 000, 6000 and 3000 gigalitres per year for the higher scenario, the base case and the lower scenario respectively. The requirement for the higher population scenario is six

Figure 8.9
Simulated urban water requirement to 2050 in gigalitres (10^9 litres) per year , for three population scenarios: the base case of 70 000 net immigration per year (70kpa), zero net immigration per year (zero) and 0.67 per cent of current population as net immigration per year (0.67%pa).
SOURCE Foran and Poldy 2000

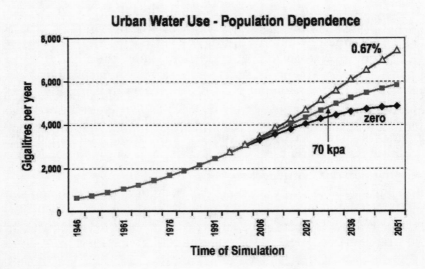

times that of current urban consumption (2000 gigalitres per year) and approximately half of current total Australian consumption (22 000 gigalitres per year). Yet the reality is that the high value of urban water would ensure that extra dams, interbasin transfers and pipelines would supplement the current urban water supplies. Thus the problem is one of allocation of available water supply, rather an absence of water. The problem therefore enters the preserve of economics and politics and moves outside the sphere of physical analysis. The modelling framework has served to quantify the size and nature of the problem and solutions are more social and political than physical.

STRATEGY FOR NATIONAL POLICY INFLUENCE

STRATEGIC PLAN

The strategic plan for the population-development-environment project, within which the OzEcco and ASFF models are used, has three linked goals. The first goal is to underpin the debate for transforming the physical economy to more sustainable modes of operation, the dematerialised, Factor 4 or Factor 10 economy as detailed by von Weizsacker et al. (1997), Ayers (1998) and others (a Factor 4 economy aims to halve energy and material usage while doubling dollar productivity). The second goal is to have accepted at national policy levels the concepts of physical analysis of the national economic structures and functions. The third goal is to contribute to changing national policy on a number of key aspects that relate to the physical economy.

The route to achieving these goals is a complex and difficult one with two important considerations. The first is the dominance of economic analysis in national assessments of population-development issues, and the belief that market mechanisms will deal with environmental problems when they are sufficiently important to require a solution. Allied with these economic beliefs is the belief that technological innovation is a driver of progress in its own right that will bypass those functions of the physical economy that require change. The second consideration is that the integration and modelling within these physical economy models is a challenging one where scientific proof in a traditional sense is difficult. In addition a modelling framework is always open to improvement. In a project management sense this can result in an imbalance between investment into modelling the outputs from scenario simulations and subsequent contributions to long-term analysis of national policy issues.

Given these constraints, the route chosen to these strategic goals is a partial and iterative one with an overall integration phase in the final two years of the process. Although the insights into strategic policy require an analysis of the whole physical economy, the business plan requires that 20 important sectors such as agriculture, building, manufacturing and energy are each investigated in a partial sense for an identified client who will underwrite the task. This allows deeper scrutiny, and appropriate model development for each main sector with a client with whom we might learn, and who might change policy directions or management on the basis of the consultations and the research report presented. The base case scenario is also further developed in an iterative manner, with additional insights from the client and the analysis undertaken. It could take ten years or longer to work through the 20 important sectors of the physical economy. Important insights of a generic or paradigm shifting nature might be lost in a welter of detail, or simply not recognised.

ANALYSES UNDER WAY

The strategic plan is currently being implemented with three main tasks under way. The first is an overarching one on the infrastructure requirements and environmental loadings that result from three population scenarios out to the year 2050. The second project involves the testing of alternative scenarios for land and water use out to 2050 and beyond, with the aim of maintaining national agricultural productivity, export income, food security and the ecological integrity of managed landscapes. The third project aims to develop scenarios for the management of Australia's marine fishery resources, and to explore the linkages between growth in population and tourism, domestic and export demand for fisheries product with the marine resources in the south-east region of Australia. The fourth part of the portfolio, currently without client and funding, covers the energy metabolism of Australia, particularly the use of fossil energy resources and subsequent greenhouse gas emissions. Most of the work is of a government or quasi-government nature and direct linkage with industry remains elusive. Attracting business clients requires a focus on seeking future opportunities rather than seeking solutions for perceived problems. In the next year an analysis of the top 100 companies will attempt to match the analytical capability of the project with strategic directions of suitably orientated companies.

DISCUSSION

THE APPROACH

This chapter has described a design approach to the opportunities and problems around future population-development-environment issues for Australia's physical economy. The approach aims to identify long-term inconsistencies in national plans, assist in searching for plans that do not have inconsistencies and to display the consequences of any plan or scenario on stocks of infrastructure and natural resources.

The approach relies on three key criteria if the project is to achieve its stated goal of influencing national policy directions on issues that relate to physical sustainability in the long-term. The first criterion is that policy makers should be active learners within the analytical process. Central to the modelling approaches is that of the analyst or user being the human dimension within the modelling procedure, rather than being the value free controller outside the modelling process. The second criterion concerns the understanding of the physical economy and its relationship to the monetary economy. The physical economy represents the vast array of physical transactions that underpin the flow of dollars and must obey the laws of thermodynamics, material mass balance and so on. The monetary economy is open to a wider array of innovation, beliefs and presumed behaviour. Both the monetary and the physical views of the economy are valid, and both should inform national policy making together.

The third criterion concerns the nature of predictive analyses versus the nature of scenario and options testing. The scenario analyses from this approach rely on a wide array of expert opinion and data analyses. These help set policy or control variables that drive simulation outcomes in a transparent and explicit manner. A simulation of a scenario may seek to test the physical feasibility of a particular ideology. Alternatively it may seek to design the pathways along which an ideology must progress if it is to attain an explicit goal by a future point in time. The preference for predictions in Australian policy circles as opposed to scenario design and testing is rooted in the world views and analytical methodologies of the four quadrants presented in the introduction. The degree to which policy analysts think of themselves as a passenger or a driver in the nation's affairs is an important distinction. A passenger may choose to predict and make policy changes at the margin. A driver may seek to re-design and foster the forces of fundamental change that are central to the concept of long-term sustainability.

ADVANTAGES AND DISADVANTAGES OF THE PHYSICAL MODELLING APPROACH

The process of model building, which combines the design of how the physical economy functions, with the data that describe that functioning, represents the key advantages for integrated physical models. Within this concept are reasonably complex calibration and validation procedures that set a valid foundation for the model in the historical period before the scenario is run to future times. These procedures enable a proof of concept to be displayed and an acceptance gained that the underlying modelling procedures that use historical data do compute appropriately. The treatment of stocks of people, cars, houses, agricultural fields and so on, is central to the concepts of momentum and inertia within the physical economy that was introduced with the four quadrants of world view presented in Figure 8.1. The dimension of stocks is not normally dealt with by most forms of economic analysis yet is central to the concept of environmental sustainability. The associated concept of physical realities within the production process is also vital and usually not included in economic models.

The modular and stepwise nature of model design and computation procedure allows relatively quick and easy partial simulations to be undertaken, and for further model development to be undertaken on a part, without disturbing the integrity of the whole. The level of detail is reasonably flexible and ranges in the ASFF model from 58 regions for agricultural productivity to 16 regions for human population dynamics to eight city airsheds for vehicle emissions and one national account for balance of trade computation. One national and international advantage, seemingly a strange one, is that a limited amount of simulation modelling of physical economies has been undertaken in a policy context, when compared to econometric modelling. This provides a possible advantage in the policy marketplace for concepts and analyses pertaining to physical sustainability. However there is little historical precedent in the promotion and refutation of integrative theories that deal with population-development-environment linkages and concepts.

The size and complexity of the analytical undertaking present an immediate disadvantage to scientific management, funding agencies, national policy analysts and scientific colleagues. The gulf between the constrained boundaries and reputable sureness of traditional reductionist research approaches, and a nationally scaled modelling approach that uses scenarios, has never been greater. Lutz (1994) noted the challenge of population-development-environment modelling in being able to combine a 'hard-wired model which only includes unambiguous relationships on which scientific consensus can be expected' with 'the soft model which can quantify all kinds of feedbacks and interactions that the user wants to define'.

This approach in design and implementation appears to be meeting these philosophical goals. However the absence of price mechanisms in both the OzEcco and the ASFF models that equilibrate shorter-term imbalances between supply and demand could pose a significant barrier to acceptance by national policy makers. Some viewpoints suggest that the physical and the economic approaches should be hybridised and blended, whereas others are satisfied to keep them as distinct and separate analytical approaches. Both modelling approaches concur that prices and market mechanisms are critical to balancing the economic concepts of supply and demand in the short term. However the strategic intent of the long-term physical modelling approach is to provide an information flow from longer-term horizons to current market, policy and business agendas.

FUTURE DESIGNS AND POLICY INSIGHTS

Once distilled, future design criteria and policy insights no longer seem particularly innovative. So it is with initial distillations from the many partial analyses so far performed with the design approach that these modelling frameworks model. Before the current work on future population targets and with a view to 2051, the lower population targets produced by zero net immigration may have seemed preferable since they stabilised a wide range of environmental loadings such as vehicle emissions. However with the 2101 view the medium population scenario produced by 70 000 net immigration seems preferable since it avoids a rapid decline in total population and the available workforce later in the 21st century and beyond. A societal and policy requirement to balance non-environmental with environmental criteria was the first insight gleaned. With hindsight this is an obvious insight, and does not require the full implementation of the ASFF model to allow its distillation.

Within a stabilising population, the design challenge is to seek technological and behavioural changes that rapidly stabilise environmental loadings and then decrease them. Unfortunately the age profile of most big ticket infrastructural items dictates that many areas of environmental pressures might continue to trend upwards for at least the next human generation. This may be so for vehicle emissions where increasing car ownership and usage is possible for the next 20 years or more, before a stabilising human population causes energy use and subsequent emissions to plateau and then slowly decline. The rapid penetration of new car technologies with much reduced energy usage may be limited by a relatively saturated vehicle ownership and a relatively old car fleet that turns over slowly. Combined with these factors is a market demand for larger more powerful vehicles, the use of which balances out the energy use by smaller more energy-efficient vehicles. Consumer behaviour may continue to keep pace with technological innovation, potentially giving a neutral outcome for any potential to decrease resource usage.

As the analyses and policy interactions proceed, the design task for the next generation of physical economy seems to become more skewed. Simple solutions to resource use and environmental loadings such as behavioural change and reducing personal consumption levels are quickly deemed less acceptable because of the flow-on effects on the monetary economy. The technological challenges then become more difficult as the redesign of the physical economy evolves to also include the redesign of the monetary economy and the social system. While this chapter describes a modelling approach centred on the physical economy, it does not negate the importance of the monetary economy. Rather it seeks to ground financial and monetary viewpoints in physical reality. However, the physical concepts underpinning sustainability suggest that a revolution might be required. When this revolution does occur, the economic structure, our personal behaviour and environmental technology will all have to move towards new configurations.

CONCLUSIONS

The concept of sustainability, linkages between energy use and greenhouse politics, population policy and lifestyle options and are all linked to environmental quality in the long term. That is not to say that larger populations live in a less sustainable manner than smaller populations. Nor does it assume that technology will find a way to overcome all environmental challenges or constraints to resource use. This chapter makes three key points. Firstly, sustainability and all the issues therein are long-term ones. Secondly, long-term issues must be explored with long-term methods that enumerate slow moving variables such as population momentum and infrastructure inertia. Thirdly, an engagement process must take place whereby decision makers are comfortable with long-term 'beyond the horizon' analyses and accept that such analyses are a valid and necessary part of the national policy process.

In order to examine the long-term consequences of many policy interactions, analytical frameworks are required to design and test different functions and structures for the physical economy. The term 'physical economy' has been coined to describe the vast array of physical transactions that underpin the monetary economy. For every dollar that is exchanged in Australia's gross domestic product, there is a chain of physical actions that brings that final good or service to the shopkeeper's counter and the consumer's basket. The processes that run the physical economy in Australia require that over 170 tonnes of material are moved per person per year to supply our essentials, our lifestyle and the exports needed to pay for our imports. By contrast, Japan moves around 40 tonnes per person while the United States moves around 80 tonnes per person.

The ability to analyse these transactions is described within two analytical frameworks, the Australian Stocks and Flows Framework and the OzEcco embodied energy flows model. The first (ASFF) is a set of 30 linked calculators that follow, and account for, the important physical actions that underpin our everyday life. The second (OzEcco) is based on the concept of embodied energy, the chain of energy flows from oil well and coal mine that eventually are included or embodied in every good and service in both the domestic and export part of our economy. Both analytical frameworks are based on systems theory and implemented in a dynamic rather than an equilibrium approach. This allows transition pathways towards new states of the physical economy to be designed and tested for physical feasibility using concepts of age and inertia. These concepts are critical to the process of infrastructure renewal, and market penetration by new technologies. The concept of physical feasibility is an important definitional idea and should not be interpreted to also reflect feasibility in a political, social or an economic sense.

The range of concepts and methodologies that are important in analysing the physical economy are described. One concept is that analysis (of the physical economy) and ideology (of the policy analyst and decision maker) should remain separated in analytical terms. The approach uses qualitative scenarios where policy analysts can design or foresee what the physical economy might be, or should be, some time in the future. Formulating these scenarios requires that assumptions be made about the likely trend of key physical parameters well into the future. Since all of these assumptions have some physical manifestation (for example, household size, human diet, engine efficiency, crop yields) they are open to scrutiny and debate. The quantitative framework that tests the summation of effects between these assumptions uses physical equations, life cycle analysis and the laws of thermodynamics to ensure that assumptions do not depart from physical reality. Many economic approaches lack this reality check.

While these approaches are relatively novel in Australian and international policy terms, the concepts therein are slowly gaining traction in parallel to a range of policy debates that are underpinned by physical realities. Energy and greenhouse, land degradation and river salinity, population growth and air emissions, oil depletion and transportation systems all represent physical realities with slow moving response times to policy interventions. Current use of the modelling frameworks is focused on long-term population policy, land and water futures, fisheries management and the decarbonisation of the transport fuels cycle. Central to the use of the frameworks is their use with, and for, clients and stakeholders. The understanding of the physical issues involved is a vital precursor to the acceptance of the radical redesigns of Australia's physical economy that might be required if the concepts behind sustainability are to be eventually implemented.

9
INFORMING INSTITUTIONS AND POLICIES
STEPHEN DOVERS

INTRODUCTION

> States should cooperate to strengthen endogenous capacity-building for sustainable development by improving scientific understanding through exchanges of scientific and technological knowledge ... each individual shall have appropriate access to information concerning the environment that is held by public authorities, including information on hazardous materials and activities in their communities, and the opportunity to participate in decision-making processes. States shall facilitate and encourage public awareness and participation by making information widely available. *Rio Declaration on Environment and Development, principles 9 and 10* (United Nations 1992, 10).

> Challenge: to improve the collection, coordination and dissemination of natural resource information and environmental information and of data systems. Strategic approach: efforts will focus on improving data collections and coordination, maximising the availability and use of existing data and activities, clearly identifying user needs and coordinating activities between different levels of government to avoid overlap and duplication. *National Strategy for Ecologically Sustainable Development, section 14* (Commonwealth of Australia 1992, 62).

These statements formed part of the response in the early 1990s to the emerging agenda of sustainable development. Although having deeper historical roots, this agenda was consolidated by the World Commission on Environment and Development (1987), and a response articulated through the 1992 UN Conference on Environment and Development (United Nations 1992), and through numerous national and sub-national processes. Australia's ecologically sustainable development (ESD) process was one of the latter (Hamilton and Throsby 1998). Since then we have grappled with the intellectual, scientific, practical and policy implications of ESD. This search for innovative, effective responses to complex, interrelated policy and management problems pervaded by uncertainty is the context of this book, especially how we *inform* our policy choices. As the statements above show, information is considered crucial; in pure scientific terms, as an input to policy and management, and as a resource for firms, communities and individuals.

We fondly hope that we choose policies in an informed manner, but Australian resource and environmental management has often been characterised by 'policy ad hocery and amnesia' (Dovers 1995; see also Toyne 1994, Walker 1994). There have been too many fits-and-starts and short-lived policies and programs, and basic information needs addressed patchily across both space and time. Even if only partly true, this is a key point for this book, for information systems, monitoring, reporting and the use of these is a prime antidote to ad hocery and amnesia. Water reviews not repeated, land degradation surveys lapsed, discontinued representative basin programs, never-realised national biodiversity monitoring systems, on-again off-again state of environment reporting, even diminished stream flow and weather recording — the list goes on. The sorts of approaches and initiatives covered in the preceding chapters of this book can serve to change this, but only if resourced, maintained and used.

The search for responses is an exercise in policy instrument choice: we have a range of interrelated problems, and must choose the response suited to each. Information systems are one of many classes of policy instrument (statute law, negotiation, agreements, education, market mechanisms, and so on). Moreover, the 'information' class contains many specific options. That 'information is important' is both true and unhelpful. Table 9.1 indicates available policy tools and suggests selection criteria specific to ESD. Instrument class 2 (communication and information flow) is the topic here, but is difficult at times to separate from classes 1 (R&D and monitoring) and 3 (educative). But all other classes relate to class 2: policy formulation requires information inputs and outputs. There are *universal* policy instruments in Table 9.1 that will *always* be required: information and communication, a basis in statute or common law, and institutional arrangements. We want policy to be informed, people need to know what is going on, we are subject to the rule of law, and we act collectively through institutions.

Table 9.1
Policy instruments for ESD; criteria* for instrument choice

Instrument class	Main instruments and approaches
1 R&D, Monitoring	Increase knowledge generally (basic research) or about a specific matter (applied research); establish a standard; develop technologies or practices; establish socio-economic implications; monitor environmental conditions or policy impact.
2 Communication and Information Flow	*Directions*: research findings to policy; policy imperatives to research; both to firms, agencies and individuals. *Mechanisms*: state of the environment reporting; natural resource accounting; community-based monitoring; corporate reporting; environmental auditing; strategic impact assessment; fora for consultation or policy debate; annual reports, etc; freedom of information.
3 Education and Training	Public education (moral suasion); targeted education; formal education (schools, higher education); training (skills development); education regarding other instruments.
4 Consultative	Mediation; negotiation; dispute resolution; inclusive institutions and processes.
5 Agreements, Conventions	Intergovernmental agreements/policies (international or within federations); memoranda of understanding; conventions and treaties.
6 Statutory	New statutes or regulations under existing law to: create institutions; establish statutory objects and agency responsibilities; set aside land for particular uses; land use planning; development control; enforce standards; prohibit practices.
7 Common Law	Torts, nuisance, public trust.
8 Covenants	Conservation agreements tied to property title.
9 Assessment Procedures	Review of effects; EIA; social impact assessment; cumulative impact assessment; risk assessment; life cycle assessment; statutory monitoring requirements.
10 Self-regulation	Codes of practice, codes of ethics, professional standards.
12 Community Involvement	Participation in policy formulation; community based monitoring; community implementation of programs; co-operative management; community management.
13 Market Mechanisms	Input/output taxes/charges; use charges; subsidies; rebates; penalties; tradeable emission permits/use quotas; tradeable property/resource rights; performance bonds; deposit refunds.

14 Institutional Change	To enable other instruments or policy and management generally, esp. over time.
15 Change Other Policies	Distorting subsidies, conflicting policies or statutory objects.
16 Reasoned Inaction	(Where justified by due consideration.)

*Criteria for instrument choice:

1. *Effectiveness criteria*: information requirements; dependability (re: goals); corrective vs antidotal focus; flexibility (across contexts, time); gross cost; efficiency (relative to achieving goal); cross-sectoral influence.

2. *Implementation criteria*: equity impacts; political/social feasibility; legal/constitutional feasibility; institutional feasibility; monitoring requirements; enforcability/avoidability; communicability (re: those affected).

But instrument choice often comprises crude advocacy of loose options, defined by ideological or disciplinary bias, convenience or bureaucratic turpitude. We can to do better. For example, with information-based policy options, if information is important, what particular strategy is required, or what mix of strategies? How do we analyse the informational need? Which of the many possible responses is best suited to the problem? Who are the users, and what are their needs? Who will pay the costs of gathering, manipulation and dissemination? Exactly what difference is information expected to make to policy? Do we know what difference information has made in the past or in comparable situations? Is there a monitoring system in place? When does data gathering equal a diversionary strategy to excuse not doing what we know we should? Do we have appropriate institutional arrangements to support monitoring and information flow?

The first part of this chapter delves behind the statement 'information is important' to suggest ways to answer these questions in different contexts, and especially the last question concerning institutions. It considers the nature of sustainability problems, and the informational nature of different policy options. The rest of the chapter considers institutional design to support monitoring, information and communication systems.

SUSTAINABILITY AND POLICY: ISSUES OF INFORMATION AND UNCERTAINTY

The issues dealt with in this collection — policy and management problems in sustainability — have attributes rendering them particularly difficult. Indeed, *different in kind and degree* from other policy fields (Dovers 1997). Unless responses are designed with regard to these attributes they are unlikely to be effective. For information-based responses, key amongst these attributes are:

• The non-market nature of many environmental assets, in a world where commodities and services traded in formal markets are generally the only ones regularly measured. This demands new monitoring and measurement techniques and processes and sometimes the creation of markets. Creating reliable information streams is difficult when policy and property rights and responsibilities are weakly defined.

- Spatial scales cut across jurisdictions, necessitating inter-governmental approaches to data gathering, analysis and use. In a federal system this problem is compounded. Also, many sustainability problems, such as biodiversity and land and water resources, cut across portfolios and sectors, demanding inter-agency and cross-sectoral co-ordination.

- The temporal scales of problems vary, but are often much longer than political and economic time horizons. Natural systems and the impacts of human interventions stretch over decades at least.[1] Policy and information processes determined by political and economic time are typically myopic from an ecological perspective.

- Demands for community participation in environmental policy are increasing. Environmental monitoring or the application of information in management will not work without the involvement of communities and landholders. However, when and how to most effectively engage communities is still not clear.

- 'Uncertainty' pervades problems in sustainability, regarding the state of natural systems, their future conditions, the impact of human activities and the efficacy of policy interventions. This attribute defines the needs for information.

Temporal scale, participation and uncertainty are crucial in a policy and institutional sense. Institutions must create processes capable of being *persistent* over long time periods and of being *inclusive* of a wide range of interests. Most critical is uncertainty — without uncertainty the issue of information would be uninteresting. Yet it is poorly handled in much environmental management and requires more exploration. We can address three common fallacies: that with enough research and monitoring things are knowable; that 'scientific uncertainty' is the only issue; and that policies and management decisions are rational and objective.

It is easy to dispose of the first fallacy. For any significant sustainability problem, it is doubtful that we will ever enjoy a semblance of certainty. Thoroughly *reducible* uncertainty is the exception. Most often, uncertainty will be *irreducible*, especially in the time span within which we will need to act. We should not avoid decisions while entrusting all to further research. This is not to say that scientific research and especially long-term ecological research and monitoring are not important — they are and we are failing in that regard. The point is to recognise uncertainty explicitly and the contingent nature of our decisions and understanding, and to improve decision making in the face of uncertainty.[2] A rough typology is useful (Dovers et al. 1996, see also Wynne 1992):

- *Risk* is where sufficient information exists for believable probability distributions to be assigned to possible outcomes or future states; we know the odds. For example, the risks associated with release of a known quantity of a well-understood pollutant into a waterway where hydrological processes and biota are well documented.

- *Uncertainty* is where, although we may be confident of the direction of likely change, we cannot assign probability distributions to future states. The general state of understanding of global climate change would fall into this category.

- *Ignorance* is where we cannot be confident of the direction of likely change, and where threshold effects and likely surprises lurk. Examples are regional precipitation impacts of climate change, or biodiversity impacts of release of genetically modified organisms. Surprise is common in the history of human-natural system interactions.

The second fallacy is more profound. Missing information or unexplained processes — scientific uncertainty — have received most attention. But there are other, important forms. Following Smithson (1989) we can use the term *ignorance*, divided into *error* (to be ignorant of) and *irrelevance* (to ignore). Within these, objective (scientific) uncertainty and quantifiable (probabilistic) risk must be considered along with perceived irrelevance, intentional distortion, concealment, confusion, taboo, surprise and other forms of ignorance. To resist such a definition would require a belief that policy debates are entirely rational and policy actors never distort information. Only one kind or degree of uncertainty will rarely attend an issue and different players in debates will not have shared understanding or information. It is an important role for information and monitoring systems to assist in reconciling these differences. The designers of such systems must be sensitive to the political nature of information and uncertainty.

The third fallacy begs a discursion on the nature of, and the role of information in, policy and politics. Decisions are not always or even often made in a 'rational' manner, based on 'objective' information. As much as scientists may not like the idea, all decisions are political and in sustainability, where conflicting ecological, social and economic information and values must be taken into account, they are doubly so. Even when more information is available political decisions are required. The Coronation Hill decision, based on the Resource Assessment Commission's detailed ecological, socio-cultural and economic investigations is an example. The balance of which of the three should dominate was rightly *political*. The then Prime Minister, Bob Hawke, emphasised cultural considerations in driving the decision not to allow the mine to proceed (Stewart and McColl 1994). The question is how information can be inserted into policy debates, sourced from what kind of research and monitoring, presented in what forms, and mediated through what kinds of institutions and processes?

Central is the definition of *uses and users* of information. There is often confusion in state of environment (SoE) reporting, corporate environmental reports and natural resource accounting. Who are the 'users' (the public, managers, politicians, the media, shareholders, etc.) and how do they use the information (general education, monitoring change, reviewing policy, informing specific decisions)? Often, SoE reports fall between these stools. Central also are the means to use limited information to support decisions and policies. Pervasive uncertainty demands explicit recognition, and a very broad suite of approaches to support decisions in conditions is demanded by multiple forms of uncertainty. Table 9.2 provides a sample of methods and techniques for informing policy in the face of uncertainty. All have information inputs and outputs. All are useful in some circumstances and reliance on only one is unwise. As well as recognising different approaches we must recognise the expertise required to apply them — economics, ecology, law, engineering, public administration, politics, accounting, psychology, and so on. The challenge lies in matching approaches to specific problems.

Table 9.2
Approaches to support policy making in the face of uncertainty*

Long-term ecological research and monitoring
Policy monitoring and evaluation
Data extension and inference through modelling
Research and monitoring of human systems
Reporting and communication systems (for example, SoE)
Quantitative risk assessment
Environmental/ecological risk assessment
Strategic risk assessment**
Extended cost-benefit analysis
Environmental impact assessment
Strategic environmental assessment
Commissions of inquiries, etc.
Regret criteria (maximax, minimax)
Safe minimum standards
Non-market valuation (hedonic pricing, contingent valuation, etc)
Discursive methods (for example, citizens' juries)
Mediation and negotiation
Community participation in policy formulation
Precautionary principle (in statute law, or as policy guideline)
Performance assurance bonds
Various planning approaches
Population viability analysis
No-regrets policy options
Adaptive management

*including analytical and evaluative techniques, legal notions, political strategies, etc.
**on risk assessment, see AS/NZS 4360 (1999 revised edition) Risk Management

If information and monitoring systems are to indicate more sustainable practices, the institutions supporting them must reflect the nature of uncertainty. Decision makers and other policy actors must achieve the difficult balance of moving forward *purposefully*, and having the humility to be *flexible* and to learn and adapt as understanding evolves.

INSTITUTIONS

Institutions are the means by which we collectively pursue goals. Institutions may be formal or informal, social or economic, small or large. Properly, *institutions* are underlying and long-lasting rules, patterns of behaviour, structures, and so on, and *organisations* the more immediate manifestations of these (for example, Henningham 1995, Goodin 1996). I will conflate the two, however, and consider organisations along with institutions on the proviso that organisations need wide recognition and some longevity to be considered 'institutionalised'. Institutional arrangements affecting the pursuit of sustainability are complex, and include legal bodies and processes, Commonwealth, state and local government agencies and processes both parliamentary and bureaucratic, international bodies, agreements and law, and industry and community organisations. Preceding chapters identify many of these. How to discuss what institutional arrangements we need, whether through analysing existing or suggesting future ones? This begs a means of describing institutions. Table 9.3 defines *attributes* of institutions — 'neutral design features'. This allows description and assessment of how an institution matches the policy and management problems it will face.

Table 9.3

Attributes of institutions (neutral design features)

SOURCE Dovers and Mobbs 1997

Extent or limits in geographical space (spatial scale)
Jurisdictional, political and administrative boundaries
Degree of permanence and longevity
Intended or actual roles, and sectoral or issue coverage/focus
Nature and source of aims and mandate
    (in custom, or statute or common law)
Degree of autonomy
Accountability (how, to whom)
Formality or informality of operation
Political nature and support (actual, required)
Exclusiveness/inclusiveness (membership, representativeness)
Degree of community awareness and acceptance
Degree of functional and organisational flexibility
Resource requirements (financial, human, material)
Information requirements (internal, external)
Reliance on and linkages with other institutions

For institutions to inform sustainability policy, from the list of approaches to handling uncertainty in Table 9.2 one is particularly relevant: 'adaptive management' (for example, Gunderson et al. 1995). This explicitly recognises uncertainty and complexity, and frames policy and management interventions as experiments aimed at improving understanding over time. This emphasises moving forward with purpose, but at the same time learning and adapting. Such an approach to policy processes and institutions is particularly relevant to this book — policy as a *learning* and *informing* system. The broad characteristics of 'adaptive' institutions and policy processes are:

- *Persistence*, where initiatives and efforts are maintained long enough for lessons to be accrued and improvements made, and institutions and processes have sufficient longevity. Impatience and expediency poison learning.

- *Purposefulness*, where policy is underpinned by goals and principles, so that firm direction is possible. Whatever their deficiencies in current expression, Australia should persevere with ESD principles given the lack of other possibilities and their wide expression in over 120 statutes (Stein 2000).

- *Information-richness and sensitivity*, given that learning and improvement are impossible otherwise.

- *Inclusiveness*, in the face of strong demands and justification for community participation in policy, management, R&D and monitoring.[3]

- *Flexibility*, to prevent persistence and purposefulness becoming rigidity. Information and learning require flexibility and the preparedness to change and adapt.

While not mutually exclusive, these five characteristics contain a number of tensions, such as between persistence and flexibility and inclusiveness and purposefulness, and these tensions should be clearly recognised.

Persistence needs emphasis, as it is relevant to policy processes and institutions underpinning information and monitoring systems — to provide mandates and maintain efforts. It also relates to the continuity and usefulness of information itself. Imperfect but useable data sets with decent time series are preferable to technically impressive but discontinuous data sets or ones otherwise unable to reflect trends over time. That is the shame of non-repeated national land degradation or water surveys, discontinued stream flow records, or fauna surveys for environmental impact statements unconnected to other data sets.

THE CHANGING 'OPERATING ENVIRONMENT' OF ENVIRONMENTAL INSTITUTIONS

Institutions have a degree of longevity, but change constantly. Creation afresh is rare, but refashioning is more common, with incremental changes occasionally disrupted by major spasms. The institutions of Australian policy have changed in response to environmental concerns, but not the significant change that many believe must come. Change is often a reaction to exogenous forces and it pays to recognise these. Four underlying political and social trends have affected policy and institutions in the past two decades, and will continue to dictate the direction of resource and environmental management. These trends are marketisation, community participation, globalisation and information technology. While related they are discussed separately here.

The first is neo-liberal political and economic ideology and its manifestations in 'marketisation' (crudely termed 'economic rationalism' in Australia). This has resulted in a growth in interest in market instruments (Table 9.1, class 13) and, more importantly, market-oriented reform of public institutions (Eckersley 1995, Dovers and Gullett 1999). This has come in the form of privatisation, corporatisation, managerialism, contracting out, outsourcing and downsizing. Major changes have occurred across natural resource and environmental management and affected how information needs are defined, how information is gathered, who owns it, who has access to it, and at what price.

The second is rising theoretical and practical interest in participatory modes of politics, policy and management. The growth of 'community-based' environmental programs and groups is one result. This trend can be at odds with marketisation, as citizens are recast as consumers. For environmental information systems, there are more players involved in policy debates, potentially more gatherers and users of information, and a widened 'peer community' judging the quality of information.

The third trend, 'globalisation', includes internationalisation and increasing uniformity in information and financial systems, corporate structures and policy and law across national boundaries. Most attention is paid to the internationalisation of finance and business. But the internationalisation of policy and law is important, where many information processes follow international standards on reporting protocols (for example, UN Statistical Office, OECD, ISO) and are used to meet reporting obligations under international instruments (for example, Framework Convention on Climate Change, Montreal process on sustainable forestry).

The fourth is the growth in information-based technologies and processes (the 'IT revolution'). The capacity, computational power and dissemination potential of IT open up new horizons of information gathering, storage and use, although whether these will be used well is not clear (see above on the use of information in policy). These changes in technological capacity and the rapid uptake of them have implications for public participation. There is the potential for rapid, wide dissemination of information, depending on access and techno-logical literacy. But, in combination with marketisation and institutional change, there is potential for arcane languages, user-pays barriers, bureaucratic secrecy and commercial-in-confidence rules to exclude people. There are questions about the long-term availability of environmental data (and policy documents) that are primarily available and archived electronically.

These trends need to be accounted for in assessing current and future endeavours. Given the nature of information and uncertainty, the importance of adaptive institutions, and the trends above, we can now consider two poorly attended information needs: ecological monitoring, and policy monitoring.

LONG-TERM ECOLOGICAL RESEARCH AND MONITORING

It is a commonplace that we understand too little about ecological systems — organisms, their frequency and abundance, life histories and ecological functions, the role of key cycles (nutrients, etc), and their vulnerability and responses to human-induced disturbances (for example, Wilson 1993, SEAC 1996, Yencken and Wilkinson 2000). For informing policy, this is the foundation. Endless information systems, indicators and reports will be useless if not supported by long-term ecological research and monitoring (LTERM). The attributes of sustainability problems — uncertainty, complexity and temporal scale — suggest that substantial, long-term investment in LTERM would be evident. I would not suggest that nothing has been done; on the contrary much has, but LTERM remains patchy and poorly co-ordinated. Ecology is not a stable, mature area of inquiry (Peters 1991, Schrader-Frechette 1995, Dovers et al. 1996). Theoretical and methodological developments are frequent and any sustained empirical investigation will provide insights into system behaviour. The stretching of policy and management across portfolios, regions and jurisdictions has seen increased emphasis in ecology on integrative, whole-of-landscape approaches — a more difficult task than traditional plot-scale investigations. The need for basic data and better predictive capacity, along with the critical nature of the issues (biodiversity, land and water, catchment management), present an area promising good payoff in scientific and policy terms.

What is not commonplace is evidence that our R&D and monitoring reflect the nature of the task. In the early 1990s, in response to emphasis on biodiversity and monitoring in policy (for example, Article 7, Convention on Biological Diversity), a national program of long-term biodiversity monitoring was proposed (Redhead et al. 1993, Aquatech 1995) but failed to proceed — another tale of unfinished business. Many have emphasised the problem but little gets done (for example, SEAC 1996, Industry Commission 1998,

Productivity Commission 1999, Yencken and Wilkinson 2000).

The following defines the nature of LTERM and barriers to it. *Monitoring* is 'the continual or continuous observation or measurement of a system or components of a system ... addresses the issue of change or lack of change through time in particular places' (Redhead et al. 1993). Monitoring is often targeted at a particular concern but, given uncertainty and complexity, there is always a likelihood of serendipitous insights. We do little monitoring of sufficient detail and longevity to track changes in natural systems — of biodiversity, of vegetation change, of surface and ground waters, and so on. What we do of sufficient detail is rarely maintained over time, and what we do of sufficient longevity often lacks detail. *Long-term ecological research* (LTER) is generally more purposeful through the statement of hypotheses and problems. Large scale, 'ecosystem experiments' are of particular value (Carpenter et al. 1995). Many key insights into Australian environments have come through *sustained* research, whether driven by researchers or defined by political and developmental factors. Important examples include: the high country, through ecological investigations associated with the Snowy Mountains Scheme; the Murray-Darling Basin, through the MDB Commission and its predecessors dating back to 1915; the Great Barrier Reef, via the GBR Marine Park Authority; central Australia through the presence of CSIRO's Centre for Arid Zone Research; and in the Alligator rivers region, due to requirements placed on the Ranger uranium mine. Two examples of recent and on-going LTER in Australia concern the impacts of landscape change on faunal assemblages in the central highlands of Victoria and the southern highlands of New South Wales (Lindenmayer 2000, Lindenmayer et al. 1999). Such research requires linkages with management and investigations over decades rather than years, and suits an adaptive approach.

How to serve long-term interests when research and monitoring often reacts to short-term political or development concerns? Recognition of the potential is one prerequisite, as is forethought about later investigations. At a finer resolution, the suggestion that a national database of environmental impact statements be created, linking this myriad of short-term studies, has much to recommend it (Just et al. 1995). EIA is basic to environmental management but is by definition ad hoc, and the information content poorly consolidated over time.

What are the barriers to LTERM? Those listed below are major ones and are interrelated:

• Lack of scientific and academic kudos attached to monitoring as opposed to research. Monitoring, while viewed by many as not intellectually innovative, requires forethought, expertise and skill.

- Bureaucratic and political attitudes are less supportive of long-term initiatives. Firstly, there is reluctance to support things that will outlive the current government. Secondly, there is scant political reward in monitoring (a Ministerial press release entitled *Still watching* is unlikely to run!). Also, frequent change in policy will result in program changes that disrupt monitoring.

- Limited tenure, employment contract or funding make it hard to maintain long-term ecological studies — two or three years of a Ph.D. project is often the limit.[4]

- The cost of monitoring can be high. Even with maximum utilisation of existing sites, proper monitoring would involve the maintenance of hundreds of monitoring sites, and a cadre of regionally based, trained operatives. It would be a significant undertaking, but not out of proportion to the task.

- Lack of mechanisms to connect temporally or spatially separated studies, so that we do not fully utilise existing programs. This stems from the fragmented institutional landscape of environmental policy, but also from deficiencies in co-operation between organisations and researchers.

- Institutional changes do not always help, as roles and structures have (sometimes inadvertently) changed between the community, public and private sectors.

Recognition of barriers is the first step, and space does not permit any detailed discussion of remedies here. For science, a reconsideration of the professional kudos attached to monitoring is warranted. There is a need for cross-institutional links through a national system. This would need to be funded, probably from the Commonwealth level, and relevant scientific associations should be involved. Linking tertiary training (ecology, geomorphology and hydrology) to long-term monitoring sites can combine applied training and the needs of management agencies. University departments could enter into on-going agreements with agencies to carry out field training through sustained monitoring.

POLICY MONITORING

One form of requisite information often not collected and evaluated is that concerning policy interventions. Australia has experimented with countless policy processes, programs and instruments, but the experiment has not been well-designed or purposeful in terms of monitoring, evaluation and application of lessons arising — policy ad hocery and amnesia. This is a result of scattered policy responsibilities across jurisdictions and portfolios, unconnected efforts across disciplines and lack of institutional arrangements to encourage policy learning. Policy learning — not a well-understood art and craft (May 1992) — is particularly difficult in sustainability, due to the attributes discussed earlier. Policy evaluation is intensely political as it involves judging policy.

The 1996 national state of the environment (SoE) report failed in attending the 'R' of the pressure-state-response (PSR) reporting model; that is, to properly evaluate policies (Anderson et al. 1997). In its defence, few involved had explicit skills in policy analysis (scientists dominated the reference groups), but the available literature on Australian environmental policy was not well utilised. Government-sponsored reporting processes will always have difficulty in commenting on the efficacy of government programs.

The Productivity Commission (1999) review of the implementation of ESD policy stands as a rare, critical government-sponsored evaluation of policy. A central finding was that implementation failed the standards of 'good policy practice'. That is a disturbing charge indicating a weak policy field. The quick demise of the official reporting mechanism for the National Strategy for ESD, the Intergovernmental Committee for ESD, confirms this. Policy monitoring is also inherently difficult, given the time lags between policy or management intervention and impact on processes within natural and human systems. Clear definition of the expected impacts of policy interventions, the time scale over which they are expected to emerge, the information required to assess policy impact and the responsibility for collecting this, are strategies to account for this. Ensuring sufficient longevity in policy monitoring and evaluation exercises is another.

This raises the questions of *what* to measure and *who* should evaluate. Funding as an indicator merely tells how much money was spent and not to what effect. In the adaptive vein, goals stated as testable 'hypotheses' demand definition of the empirical evidence required to test the proposition and of the means of monitoring. Part of ad hocery is emphasis on the 'policy statement' at the expense of what comes before and after in problem definition, information, implementation and evaluation (Dovers 1995). The design of monitoring and evaluation, with responsibilities assigned, must be part of policy design from the start. In terms of who should evaluate, until the often-defensive nature of politics changes, independence is a prime requirement. Auditors-general have performed this function well at times (for example, ANAO 1997), however, generic agencies may lack familiarity with sustainability issues. Independent commissioners or commissions (the latter allow stakeholder representatives) have attractive features for tasks such as SoE reporting. Central statistical agencies are candidates too (the Productivity Commission recommended the Australian Bureau of Statistics play a lead role).

A review of institutional arrangements for policy monitoring is warranted, including of linkages across sectors, portfolios and jurisdictions. Recent institutional changes, largely driven by marketisation, make a review particularly necessary. More than a snapshot is needed

— information capacities are a long-term, fundamental need and should be regularly evaluated. At a finer resolution, there are often evaluations of specific programs by auditors-general or consultants (see Curtis et al. 1998). A critical need is to state clear program goals and performance criteria from the outset as vague goals are not testable. Furthermore, the data required to test the achievement of program goals should be routinely gathered and made easily accessible.

The institutions and policy processes required for information systems need to pay attention to these issues of ecological and policy monitoring. If not, we will attend only to the 'middle' of the story, messing about with indicators and reports without the ecological information these should be based on or knowing what our policies are achieving.

INFORMING INSTITUTIONS AND POLICIES

The preceding discussion established key features of sustainability and the operating environment and indicated broad parameters for 'informing institutions and policies'. The equivocal meaning of that phrase is intentional — all institutions need to both be informed and serve to inform. This section engages with the sorts of processes and methods canvassed in the preceding chapters. A first cut is to separate into *processes*, and *methods* and *techniques* for particular purposes. For both, the questions are: who will use the information, for what purpose and in what institutional context?

With respect to the last question, Table 9.4 defines attributes of 'adaptive' institutions and organisations, more purposively than the earlier five principles. These attributes have been distilled from the policy and institutional literature, lessons from sustainability policy and 'encouraging' institutions in Australia, and guide analysis over the next few pages. It is not possible to deal with all the processes and methods covered in this book and elsewhere. Summary comments will be made on some in terms of the questions above, and then on SoE in terms of the attributes in Table 9.4.

With regard to *techniques*, there are many and all are both contested and useful, but precisely what they are useful for deserves comment. For example, what operational decisions could be informed by determining the 'ecological footprint' of an individual, community or settlement? The data are mostly available and the exercise is unlikely to tell an informed decision maker anything new. Simpson et al. (2000) undertook such an analysis for Australia with the conclusion that Australia is a high consumption society. The analysis is credible and fresh but the conclusion unsurprising. There are three purposes of such techniques: to inform specific decisions, to inform general policy directions and to influence community values and understanding.

Table 9.4
Attributes of adaptive institutions (*purposeful* design features)
SOURCE Dovers and Mobbs 1997; Dovers and Dore 1999

Attribute	Explanation
1. Purposeful	Clearly stated guiding (ESD) principles and mandate to pursue them
2. Longevity	Sufficient time to persist, experiment, learn and adapt (including maintenance of institutional memory)
3. Participatory	Participatory structure and process that is clear, genuine, predictable and maintained over time
4. Appropriate scale	Spatial scale appropriate to the natural and human processes most closely related to the institution's purpose, with clear connection to loci of legal and administrative power
5. Resourcing	Sufficient resources to pursue and achieve goals, including human, financial, informational and intellectual/professional resources
6. Statutory basis	Statutory base providing transparency and accountability, and a higher probability of persistence (this has three dimensions: *existence* of enabling legislation, *appropriateness* of this, and *full use* of powers)
7. Independence	Degree or independence and/or removal from day-to-day political pressures, and not overly reliant on temporary mandate or resources
8. Multiple functions	Integration of research and management/policy roles
9. Applied	Degree of applied or grounded focus (region, issue or sector), to ensure practicality and 'ground truthing'
10. Integrative	Able to integrate environmental, social and economic aspects, and pursue cross-sectoral, cross-problem and/or cross-cultural views
11. Collaborative	The maintenance of linkages with organisations and processes in cognate areas — accepting that no single arrangement can integrate all aspects and therefore there is a need to work collaboratively
12. Comparative	Ability or mandate for comparative analysis (concurrent or sequential)
13. Experimental	Mandate and ability to experiment with approaches and methods, and to move across disciplinary and professional boundaries
14. Political support	Political support favouring establishment and continuation

NB attributes may be fulfilled within a specific institution or organisation, and/or through linkages and collaboration with others.

Ecological footprinting attends the last of these, by emphasising consumption and its impacts. If the use of footprinting is heuristic, it is an educative rather than scientific exercise and should be viewed as such.

For heuristic purposes, indicators need not be complex. For comparing countries' environmental performance, the UN Development Program's Human Development Index is an example (UNDP 2000). The move from straight per capita GDP to adjusted GPD combined with literacy and longevity was easy, quick and effective. Further adjusted for energy use — the most useful single indicator of environmental impact — gives a sustainability indicator (for 126 countries see Dovers 1994). In a comparable exercise, Common (1995) constructed an indicator for 132 countries from available data on GDP and greenhouse gas emissions. In both, Australia's high consumption is identified, as is the point that high energy use is not a strict precondition for human development.

People Environment Process and urban metabolic models (see Chapter 3) indicate another issue: the institutional home for such exercises. Urban metabolic approaches have been understood since the work of Boyden et al. (1981) in their Hong Kong study. Now various interests are resurrecting the idea without appreciation of previous work. Continuity over time is crucial to improvement and application. Similar is the story of population-environment modeling at regional scale in Australia. The Commonwealth's Bureau of Immigration, Multicultural and Population Research (BIMPR, previously BIR) in the early 1990s focused on environmental aspects of population change (Dovers et al. 1992, Norton et al. 1994). The Howard government reduced the BIRMP to a statistical departmental rump in 1996, ensuring that the following years of divisive debate over immigration were less informed. One of the Bureau's last projects was a regional population-environment-economy modeling exercise (Cardew and Fanning 1996). Related work never materialised. Future debates about population policy will be poorly informed without a base of public information and policy and institutional arrangements conducive to proper debate. These elements are largely missing from the population-environment debate (Dovers 1998).

Corporate environmental reporting is both a process and a range of techniques (see Chapter 6). As a process, it may be contained within a firm or across firms in a sector, part of a self-regulatory approach or mandated by government. Presently in Australia, there are a range of forms of corporate environmental disclosure and reporting and only incomplete surveys are emerging describing their extent and coverage (SMEC/AIG 2000). Corporate reporting illustrates the different users and uses of environmental information systems. Is it for shareholders to justify expenditure or for public relations purposes? Is it for internal management purposes or to fulfil regulatory requirements? To serve all purposes is difficult and demands multiple layers of information gathering and presentation.

Most public reports seem to be for the first two purposes, which is valid, but begs the question of connection to management decisions. While reporting may tell the firm that improvement is required in their operation, this may well have already been known, and certainly the information needed to redress the situation is unlikely to be part of the reporting process. It will be more detailed and quite likely sensitive. Like SoE reporting (as discussed below) it is less the reporting process that will make a difference than what it is connected to. This is difficult in many, commercially sensitive situations. The degree of information exchange that could inform improved practice between firms within sectors is unclear except where there are sector-wide strategies in place (for example, the chemical industry's Responsible Care program). The link between private and public sector environmental reporting is an area where institutional reform might improve information flows. For this reason, institutional possibilities given later in this chapter include major stakeholders (for example, industry) as well as government.

Turning to *processes*, we can consider natural resource accounting (NRA) and SoE reporting. NRA aims to embed environmental considerations more deeply in public policy (see Nordhaus and Kokkelenberg 1999). The assumption is that if the environment were part of the national accounts then it would matter more. Integrated 'green national accounting' faces difficulties in reconciling environmental and economic data that may or may not be overcome. If accounts are satellite, physical ones, the question arises of why they are not part of SoE. The assumption is questionable in its construction of the political process. It is not clear that national accounts are as influential as some think and the assumption is based on another: that economic data are all-powerful. However, governments do not always make decisions as 'rational' responses to economic data. They make political decisions based on the advice of staff and on political information concerning party policy and electoral receptiveness. Does or could economic data, whether 'greened' or not, shift political views? Perhaps the heuristic use is more important than the decision support use? Common and Norton (1994) concluded that ecological monitoring would give more useful information than adjusting national accounts.

However, NRA does at least have an independent home in a central statistical agency. The Australian Bureau of Statistics' recent forays into resource accounts can illustrate these points. The accounts for water (catalogue no. 4610.0), fish (4607.0) and energy (4604.0) do not contain much information previously unavailable to decision makers in those sectors, although they repackage it for a different audience. Analysis of the use of these accounts in *decision-making contexts* would be valuable. For heuristic purposes, it may be that broad valuations of ecosystem services (for example, Costanza et al. 1997) would be more influential than detailed accounts. Connection between accounts and other processes (for example, SoE) is unclear. Continuity with previous initiatives can be poor, as between the water account and the previous national assessments of 1965, 1975 and 1985 that reported within hydrological and water system boundaries rather than the ABS' state borders (see Dovers 1995).

SoE serves to explore the attributes of adaptive institutions (Table 9.4) and the question of uses and users. Although much is written on the purpose of SoE, it is not often clear from SoE reports. The 1996 Australian national report is a case in point. It is a useful compendium and gazetteer with implicit policy messages, rather than a policy-support tool. Little information not already available to interested policy makers was included, suggesting that the main users will be outside the policy community. It was a summary and communication exercise. Perhaps reports become more heuristic the broader the scale: the Commonwealth has little direct environmental management power. In New South Wales, for instance, there is local and state government reporting, with greater content of new or unavailable material and the value of encouraging integrated environmental considerations in local government. Ideally SoE reports would be linked across local, regional, state and national levels, allowing more or less detail at different administrative scales, and a transition from the decision-relevant to more heuristic.

SoE could be a prime response to the basic adaptive principles put earlier — purposefulness, flexibility, and information-richness. But SoE does less well against the more detailed attributes in Table 9.4. The following comments summarise key concerns against these attributes (this is indicative, not exhaustive):

- *Attribute 1: purposeful.* The idea of SoE provides for common purpose, but there is confusion as to the intent of SoE. This may be viewed as understandable given the short period we have been doing SoE (see next point). In time, explicit procedures and competencies should be developed across SoE processes and jurisdictions.

- *Attribute 2: longevity.* The whole point of SoE is for it to be ongoing. But some Australian reporting processes have been discontinued — the Commonwealth from 1985–87 and Victoria 1987–91. Currently, the longest-standing SoE processes in Australia are less than 15 years old, and there is little evidence to allow judgement as to impact (see point 5, below).

- *Attribute 3: participatory.* SoE is not an inclusive process, whether that is viewed as good or bad. Given that environmental management is increasingly reliant on self-regulatory and community-based approaches, there is a logic to ensuring participation extends to information systems. At national and state levels, should SoE be the task of government agencies responsible to the government of the day, given potential sensitivity of 'Response' indicators? Statutory independence is one answer, especially if the agency responsible is answerable to a representative board. Stronger links to the community are possible with local government.

- *Attribute 4, resourcing.* There is plentiful room for extra resources and this is widely recognised. However, resource issues (financial, human and informational) are even more crucial for the underpinning tasks of maintaining data over time. Thus, more to the point is the desirability for SoE to be the function of a dedicated agency engaged full-time — before, during and after the production of 'the report'.

- *Attribute 5, statutory basis.* Without an assigned statutory responsibility SoE processes are not fully embedded in the policy system. Too many initiatives existing in the precarious realm of 'policy initiative' have disappeared without due consideration. Sustainability is certainly considered to be important, and informing sustainability policy deserves the stronger (although not unassailable) basis of statutory recognition. In New South Wales, Queensland, South Australia, Tasmania, the ACT and at Commonwealth level, SoE has a legislative base, but not in Western Australia (Productivity Commission 1999). Victoria's lapsed SoE reporting had no legal basis and was cut swiftly by a conservative government. This is a pity, because the reporting style — a rolling, detailed sectoral review — was an interesting comparison with the more occasional, complete reports done elsewhere.

- *Attribute 6, independence.* A degree of independence from government is desirable, for greater perceived authority, continuity of function, proof from political fashion and to include stakeholder representatives in the process. The situation in Australia varies, with reports being authored solely by government, under the auspices of independent bodies, by external consultants, or a combination.

- *Attributes 7–8, multiple functions, applied.* For SoE, this can translate as needing to ensure, over time, feedback from policy and management on the effectiveness and impact of information generated through of SoE.

- *Attribute 9, applied.* If SoE is to address sustainability, rather than just the environment, it needs to integrate social, economic and ecological information. Most Australian processes do this to some extent, but failings in this regard reflect our poor abilities rather than failing of SoE *per se*. This begs the question of the use of the sorts of integrative techniques covered in preceding chapters (NRA, ecological footprints, etc.). In SoE this is subject to ongoing research and development.

- *Attribute 10, integrative.* It is not clear that SoE in any jurisdiction adequately integrates or draws information from all relevant portfolios and sectors, especially those not directly engaged in environmental management. While this is partly due to the newness of SoE it is also a reflection of the fragmented and unco-ordinated nature of the sustainability field. Further, there is little connection between SoE at different scales of reporting and between jurisdictions. At the national level, there is insufficient connection between information activities undertaken through different portfolio areas, including resource accounts (for example, Bureau of Statistics), streams of resource sector data (for example, ABARE), the National Land and Water Audit, SoE, and various sectoral policy review processes (for example, biodiversity, oceans).

- *Attribute 11, collaborative*. SoE is only one part of the information-policy continuum, and if not linked to the sources and users of information is pointless. Two tasks emerge. First of all, close links are required between primary data gathering organisations so basic data are translated into management and policy information, and policy needs communicated to monitoring agencies. Secondly, effort is needed in evaluating the *use* of SoE information in actual decision making. One determinant of this is the organisational structure of SoE processes, and the degree to which they are connected to other parts of the policy system. SoE cannot be a separate task undertaken in an undervalued bureaucratic corner.

- *Attribute 12, comparative*. This means allowing tracking trends over time (temporal), and between sectors and jurisdictions (spatial). For the first, this will take time and fulfilling attributes 2–5. For the second, attributes 8–11 need to be fulfilled.

- *Attribute 13, experimental*. This responds to uncertainty and the adaptive principle of flexibility, requiring balance between maintaining temporal consistency and ensuring improvement through change. Judgement on this will need to wait.

The section above indicates some of the major issues. One need emerging is for a policy field that is more coherent institutionally, where monitoring and information systems are developed, evaluated and evolved in a co-ordinated rather than a fragmented fashion.

TOWARDS A COHERENT POLICY AND INFORMATIONAL FIELD

As the only nation state occupying a continent, wealthy and well-educated Australia has little excuse not to have fully co-ordinated and integrated environmental information systems, including ecological and policy monitoring. Yet numerous analyses have found this is not the case. Arguments that a federal system makes this difficult are not supported by information consolidation in other policy fields (see below). Arguments that the size of the country and low population densities make for difficulties ignore the potential of remote sensing and that large areas do not require intensive or frequent monitoring. While most environmental responsibilities reside with the states, one area for the Commonwealth to play the lead role (and pay the bulk of costs) is in institutional arrangements for a co-ordinated approach. The following examples of national arrangements would help achieve this. As stressed later, such arrangements are within the normal parameters of governance and compare with arrangements in other policy fields.

- At minimum, as recommended by the 1990–92 ESD working groups but ignored by government, Offices of ESD in first ministers' departments would oversee and co-ordinate whole-of-government approaches. Part of the task would be to maximise information exchange across portfolios.

- Firm institutional arrangement for national scale information and monitoring systems and the co-ordination of state, regional, catchment and local scale systems are implicitly recommended by many analyses (for example, Productivity Commission 1999, Yencken and Wilkinson 2000). While many recommendations only deal with the Commonwealth, it would be preferable to include a national co-ordination and enabling role as well. A properly resourced National Commission for ESD under a representative board (three levels of government, science and major stakeholders) with a statutory mandate — including co-ordination of SoE reporting and carriage of national SoE — is one option.

- The R&D challenges attending sustainability demand more than the current fragmented efforts. Australia's R&D corporations are impressive (Lovett 1997). The Land and Water Research and Development Corporation has acted as the de facto ESD R&D corporation, but with a modest budget and a mandate excluding the urban, air and marine realms. A full ESDR&DC is warranted to provide strategic, co-ordinated directions in a key economic, ecological and social area over coming decades. Whether to substantially expand LWRRDC or design a new arrangement requires discussion. This could be linked, in statute and function, to the above Commission.

- Policy ad hocery and amnesia and the poor state of policy monitoring and learning deserves redress through institutional reform. Ministerial Councils, although valuable, are slow and top heavy. Many professional and bureaucratic mechanisms are poorly co-ordinated and resourced and often are contained within specific areas. There is room for a joint Commonwealth-state-local government initiative, involving professional, industry and NGO stakeholders, to serve as a mechanism for gathering, analysing and communicating policy experiences across the ESD field. An Australia Institute of ESD, based in regional Australia, would function as a clearing house for policy information, maintain a focus on emerging policy challenges, maintain an information service and run training courses. The focus would not only be on government policy, but also on industry and community experiences (for example, community monitoring, corporate environmental reporting).

- As well as policy monitoring, there is need for co-ordination across jurisdictions and sectors for basic environmental information. Existing arrangements, while serving some of our needs, have been found wanting (for example, Productivity Commission 1999). A national information gathering and clearing house facility would supplement and/or replace existing information co-ordination mechanisms. This could be linked to the above R&D and policy learning ideas.

- The functions fulfilled by the Resource Assessment Commission (RAC), especially application of innovative methods to inform policy, were valuable. Promising decision and policy support methods — many are surveyed in this book — need to be applied in realistic contexts in a transparent fashion. In its unnaturally short institutional life (1989–93), the RAC did this with contingent valuation, multi-criteria analysis and mediation approaches (for example, Stewart and McColl 1994). This could be attended to by a resurrection, or by giving that mandate to another institution.

- As well as national mechanisms, within state and Commonwealth governments, connection between different information-related activities across sectors and portfolios are needed. This applies especially to the integration of environmental, economic (production and consumption) and social (health, demographics, etc.) information to inform ESD policy. Commissioners for the Environment, Offices of ESD in first ministers' departments, or more representative statutory bodies are means of ensuring co-ordination. The forthcoming Victorian Commissioner for ESD may serve as a model for other jurisdictions.

The aim is not to supplant government. In Churchill's words, the Westminster system of parliamentary democracy is the worst system in the world except for all the others. Governments should make political decisions, but the above would enable better decisions, based on transparency, good information and the flexibility to learn. While discussed separately there are functional linkages that could be established via a single, multi-objective statute. If the 'partnership' rhetoric of recent policy is serious, then arrangements would be controlled (not just advised) by bodies representing different levels of government and key interest groups, and would have statutory independence.

This is quite an agenda and opposition would be stiff. The RAC met an untimely end due to the fact that it had an explicit sustainability mandate, was independent, reported directly to the Prime Minister and was capable of changing the nature of the debate. Not everyone wishes ESD to be a strong policy field. Lest these reforms seem fanciful, consider whether such arrangements would be unusual in other policy fields. ESD policy in Australia is weak, fragmented and poorly supported (Dovers 1995, 1999). Supposedly, we are 'integrating' ecological, social and economic considerations, but this requires institutional and informational parity between the three. There is not parity, especially between ecological and economic rationalities. To emphasise this, consider the above institutional suggestions and their low likelihood of support against comparable arrangements in other fields. Other whole-of-government issues have a formal presence in first ministers' departments, such as the arts and women's and indigenous affairs. An ESDR&DC has comparators in the other commodity-based Australian R&D corporations and in the National Health and Medical Research Council. A national policy co-ordination, learning and training institute has an impressive precedent in the (Commonwealth-funded) Australian Emergency Management Institute. A national information gathering and clearing house facility equals the Australian Institute of Health and Welfare while a resurrected RAC matches the Productivity Commission. In all these cases, a degree of statutory independence from government departments is provided.

INSTITUTIONS FOR PARTICIPATION

Emphasis has been given here to inclusiveness in higher-level arrangements, but there is a pressing issue with inclusiveness at finer resolutions of monitoring and management. For many people, the key area requiring institutional support is the range of community-based programs and groups — Everything Care and Everything Watch (Dovers 2000a). These groups, epitomised by several thousand Landcare and Waterwatch groups established in the past decade, are vastly encouraging and may be a key turning point in human-natural system interactions since European occupation. But the next few years may hold a crisis, as voluntary human resources are squeezed and falling government support feeds the belief that 'empowering communities' is code for government passing the buck. Information flow from community programs is growing and *could* be a core information strategy. But not if groups are subject to changing policy fashions, program shifts and reshuffles, and reliant on chancy annual funding rounds. Debate is required about 'institutionalising' community participation — not rigidly, but to guarantee basic resources and longevity and patterns of information ownership and flow. Community groups engaged in monitoring must know that their commitment will still be rewarded in the future, that they will have the wherewithall to undertake the work, and where the information will go and that it will be used. Basic administrative capacity (for example, a convener) should not be a short-term project. If Australian governments do not make a longer-term commitment, the enthusiasm shown by the community may turn sour, to the detriment of civic culture and the environment.

CONCLUDING COMMENT

We have come some way in informing sustainability policy and in actual policy. This book showcases advances. Australia has at times even developed world class policies, laws and management regimes. However, there is still much to do, and serious gaps remain in our information base and our attempts to rectify these. We have not evaluated past lessons sufficiently, or integrated this knowledge to inform on-going improvement. The institutional arrangements and policy processes for informing and monitoring are too fragmented. Some illustrative solutions have been given in this chapter. ESD in Australia is an institutionally weak policy field, and encouraging initiative will not bear fruit until the institutional landscape of resource and environmental management becomes stronger and more integrated.

NOTES

1   It may be that, of all human systems, it is underlying institutions that operate in time most similarly to ecological systems, with long time frames of change, but with the propensity for thresholds and sudden shifts within longer movements.

2   The notion of 'scientific certainty' is central to the precautionary principle, an ESD principle now stated or referred to in hundreds of Australian policies and statutes (see Dovers and Handmer 1999, Stewart 1999, Stein 2000). As a core element of sustainability and one irrevocably linked to available (or unavailable) information, the precautionary principle is of great relevance to information and monitoring systems.

3   Community participation means much more than the recent (and very positive) explosion of community based programs (for example, Landcare, Waterwatch). It includes legal standing in planning law, FoI, inclusion in policy formulation, electoral rights, and many other forms. This is an area that has received far too little attention (Dovers 2000a).

4   One response to this and other barriers is to make greater use of environmental history investigations to support contemporary studies through establishing baselines and time series (for example, see Dovers 2000b).

REFERENCES

CHAPTER I

Bowler, PJ (1992) *The Fontana History of the Environmental Sciences*, Fontana, London.

Carson, R (1962) *Silent Spring*, Houghton Mifflin Co., Boston.

Club of Rome (1972) *Limits to Growth*, Universe Books, New York.

Commonwealth of Australia (1986) *State of the Environment in Australia*, Australian Government Printing Service, Canberra.

Commonwealth of Australia (1992) *National Forests Policy Statement*, Australian Government Printing Service, Canberra.

Commonwealth of Australia (1992) *National Strategy for Ecologically Sustainable Development*, Australian Government Printing Service, Canberra.

Daly, HE & Cobb, J (1989) *For the Common Good: Redirecting the Economy toward Community, the Environment, and a Sustainable Future*, Beacon Press, Boston.

Deegan, C (1998) Environmental reporting in Australia: We're moving along the road, but there's still a long way to go. *Environmental and Planning Law Journal* 15(4), 246–63.

Dovers, S (1996) Processes and institutions to inform decisions in the longer term. In R Harding (ed.) *Tracking Progress: Linking Environment and Economy through Indicators and Accounting Systems*. Proceedings of the 1996 Fenner Conference on the Environment, 30 September – 3 October, University New South Wales, Sydney.

Eckersley, R (ed.) (1998) *Measuring Progress: Is Life Getting Better?* CSIRO Publishing, Melbourne.

Ehrlich, P (1968) *The Population Bomb*, Ballantine, London.

Elkington, J (1999) Wizards of Oz. *Tomorrow* July/August 1999, 54–7.

Elkington, J, Kreander, N & Fennell, S (1998) Targeting the non-reporters. *Tomorrow* September/October 1998, 58–61.

Environment Protection Authority NSW (1997) *Corporate Environmental Reporting — Why and How*, EPA NSW, Sydney.

Environment Protection Council (1986) *The State of the Environment Report for South Australia*. Department of Environment and Land Management, Adelaide.

Fayers, C (1997) *Corporate Environmental Reporting, Investment Decisions, and ESD in Australia: A Discussion*. Department of Geography and Environmental Science,

Monash University, Working Paper Number 39.

IPCC (International Panel on Climate Change) (1995) *Summary for Policymakers of the Contribution of Working Group 1 to the IPCC to the Second Assessment Report,* Cambridge University Press, Cambridge.

McCormick, J (1992) *The Global Environmental Movement: Reclaiming Paradise,* Belhaven, London.

Ponting, C (1991) A *Green History of the World,* Sinclair-Stevenson, London.

SEAC (State of the Environment Advisory Council) (eds) (1996) *Australia: State of the Environment 1996,* CSIRO Publishing, Melbourne.

UNCED (United Nations Conference on Environment and Development) (1992) *Agenda 21,* UNCED Secretariat, Geneva.

UNEP (United Nations Environment Programme) (1999) *Global Environmental Outlook 2000,* Earthscan Publications, London.

WCED (World Commission on Environment and Development) (1987) *Our Common Future,* Oxford University Press, Oxford.

WMO (World Meteorological Organisation) (1995) *Scientific Assessment of Ozone Depletion, 1994,* Global Ozone and Monitoring Project, Report No. 37.

CHAPTER 2

Attfield, R (1983) Western traditions and environmental ethics. In R Elliot & A Gare (eds) *Environmental Philosophy,* University of Queensland Press, St. Lucia, pp 201–30.

Attfield, R (1991) *The Ethics of Environmental Concern,* The University of Georgia Press, Athens, (second edition).

Bennett, DH (1996) 'When you get up in the morning': equity, Aboriginal and Torres Strait Islander peoples, native title and the environment. In M Common (ed.) *Environmental Economics Seminar Series, Equity and the Environment,* Department of Environment, Sports and Territories, Canberra, pp 49–53.

Black, J (1970) *The Dominion of Man: The Search for Ecological Responsibility,* Edinburgh University Press, Edinburgh.

Bodian, S (1982) Simple in means, rich in ends: a conversation with Arne Naess, *Ten Directions,* Summer/Fall, 7, 10–2.

Carson, R (1962) *Silent Spring,* Houghton Mifflin Co, Boston.

Coates, P (1998) *Nature: Western Attitudes since Ancient Times,* Polity Press, Cambridge.

Common, M (1996) Background paper. In M Common (ed.) *Environmental Economics Seminar Series, Equity and the Environment.* Department of Environment, Sports and Territories, Canberra, pp 6–15.

Commonwealth of Australia (1992) *National Strategy for Ecologically Sustainable Development,* Australian Government Publishing Service, Canberra.

Commonwealth of Australia (1996) *National Strategy for the Conservation of Australia's Biological Diversity.* Commonwealth Department of the Environment, Sport and Territories, Canberra.

Daly, HE (1992) *Free Trade, Sustainable Development and Growth: Some serious Contradictions,* Special Network Supplement, Reviews of Agenda 21, Network '92, The Centre for our Common Future and the IFC, Geneva, 2pp. (A Review of document number A/Conf.151/PC/100/Add.3).

Davis, T (**nd**) What Is Sustainable Development? Web site: http://environment.about.com/culture/issuescauses/environment/gi/dynamic/offsite.htm?site=http://www.menominee.com/sdi/articles/whatis.htm.

Ecologically Sustainable Development Working Groups (1991) *Final Report, November 1991.* Australian Government Publishing Service, Canberra.

Elkington, J (1998) *Cannibals with Forks.* New Society Publishers, Gabriola Island, B.C.

Five E's Unlimited (1999) Sustainable Development: De-mystifying the Concept of Sustainable Development? Web site: http://www.eeeee.net/ee00002.htm.

Hamilton, C (1996) Thinking about the future: Equity and sustainability. In M Common (ed.) *Environmental Economics Seminar Series, Equity and the Environment*. Department of Environment, Sports and Territories, Canberra, pp 16–21.

Happold, DCD (1995) The interactions between humans and mammals in Africa in relation to conservation: A review. *Biodiversity and Conservation*, 4(4), 395–414.

Harding, R (1996) *Sustainability: Principles to Practice, Outcomes*, Fenner Conference on the Environment, Canberra, 13–16 November 1994. Department of the Environment, Sport and Territories, Canberra.

Harding, R, Young, M & Fisher, E (1996) *Sustainability: Principles to Practice. Background papers*, Fenner Conference on the Environment, Canberra, 13–16 November 1994. Department of the Environment, Sport and Territories, Canberra.

Harris, S & Throsby, D (1998) The ESD Process: background, implementation and aftermath. In C Hamilton & D Throsby (eds) *The ESD Process: Evaluating a Policy Experiment*. Academy of Social Sciences in Australia and Graduate Program in Public Policy, Australian National University, Canberra, pp 1–19.

Horton, D (2000) *The Pure State of Nature: Sacred Cows, Destructive Myths and the Environment*. Allen & Unwin, Sydney.

IGAE (Intergovernmental Agreement on the Environment) (1992) May 1992, Heads of Government in Australia.

Jacobs, M (1996) *The Politics of the Real World*. Earthscan, London.

King, D (1998) A Guide to the Philosophy of Objectivism. Web site: http://www.vix.com /objectivism/Writing/DavidKing/GuideToObjectivism/index.htm.

Lebel, L & Steffen, W (eds) (1998) Global environmental change and sustainable development in Southeast Asia: Science Plan for a SARCS Integrated Study. Southeast Asian Regional Committee for START (SARCS), Canberra.

Lothian, A (1998) ESD in state government decision-making. In C Hamilton & D Throsby (eds) *The ESD Process: Evaluating a Policy Experiment*, Academy of Social Sciences in Australia and Graduate Program in Public Policy, Australian National University, Canberra, pp 53–67.

Naess, A & Sessions, G (1984) Basic principles of deep ecology, *Ecophilosophy*, VI, 3–7.

Routley, V (1975) Critical Notice (review *of Man's Responsibility for Nature* by J Passmore). *Australian Journal of Philosophy*, 53, 171–85.

SEAC (State of the Environment Advisory Committee) (1996) *Australia State of the Environment 1996*, CSIRO Publishing, Melbourne.

Simon, J & Myers, N (1994) *Scarcity or Abundance? A Debate on the Environment*, WW Norton: New York.ebook:http://www.inform.umd.edu/EdRes/Colleges/ BMGT /Faculty/JSimon/Norton/.

Singer, P (1977) *Animal Liberation*, Granada, London.

Sylvan, R & Bennett, D (1994) *The Greening of Ethics*, White Horse, Cambridge.

Tickell, C (1997). The human species: a suicidal success. In A Goudie (ed.) *The Human Impact Reader: Readings and Case Stud*ies, Blackwell, Oxford, pp 450–60.

WCED (World Commission on Environment and Development) (1987) *Our Common Future*, Oxford University Press, Melbourne.

World Conservation Strategy (1991) *Caring for the Earth: A Strategy for Sustainable Living*, IUCN/UNEP/WWF.

Yencken, D & Wilkinson, D (2000) *Resetting the Compass: Australia's Journey towards Sustainability*, CSIRO Publishing, Melbourne.

CHAPTER 3

Australian Bureau of Statistics (1996) *Australians and the Environment*, ABS, Canberra.

ALGA (Australian Local Government Association) (1996) *National Local Sustainability Survey — Occasional Paper* Environs Australia, Sydney.

Bakkes, JA, van den Born, GJ, Helder, JC, Swart, RJ, Hope, CW & Parker, JDE (1994)

An Overview of Environmental Indicators: State of the art and perspectives, United Nations Environment Programme.

Boyden, S, Millar, S, Newcombe, K & O'Neil, B (1981) *The Ecology of a City and its People: The Case of Hong Kong*, ANU Press, Canberra.

Chesson, J & Clayton, HA (1998) *Framework for Assessing Fisheries with respect to Ecologically Sustainable Development*, Bureau of Rural Sciences, Canberra.

City of Adelaide (1996) *City of Adelaide Environmental Management Plan — Local Agenda 21 Draft for Discussion*, City of Adelaide, Adelaide.

Commonwealth of Australia (2000) *A Framework for Public Environmental Reporting — An Australian Approach*, Environment Australia, Canberra.

Department of Finance (1994) *Do Evaluations: A Practical Guide*. Department of Finance, Canberra.

DETR (Department of the Environment, Transport and the Regions) (1998) *Sustainability Counts*, Department of the Environment, Transport and the Regions, United Kingdom. Web site: http://www.environment.detr.gov.uk//sustainable/consult/sust02.htm.

Global Reporting Initiative (1999) *Sustainability Reporting Guidelines — Exposure Draft for Public Comment and Testing*, Coalition for Environmentally Responsible Economies, Boston. Web site: http://www.globalreporting.org/Guidelines/Guidelines.htm.

ISO Central Secretariat (1998) *ISO 14000 — Meet the Whole Family!* ISO Central Secretariat, Geneva.

Mageau, MT, Costanza, R & Ulanowicz, RE (1995) The development and initial testing of a quantitative assessment of ecosystem health. *Ecosystem Health*, 1, 201–13.

Mann, RE (1993) Monitoring for ecosystem integrity. In S Woodley, J Kay & G Francis (eds) *Ecological Integrity and the Management of Ecosystems*, St Lucie Press, Delray Beach.

Montreal Process Implementation Group for Australia (1997) *Australia's First Approximation Report for the Montreal Process*, Pirie Printing, Canberra.

Natural Step (2000) Web site: http://www.naturalstep.org.

Newman, P, Birrel, B, Holmes, D, Mathers, C, Newton, P, Oakley, G, O'Connor, A, Walker, B, Spessa, A & Tait, D (1996) Human Settlements. In State of the Environment Advisory Council (Eds) *Australia State of the Environment 1996*, CSIRO Publishing, Melbourne.

Newman, P & Kenworthy, J (1999) *Sustainability and Cities*, Island Press, Washington DC.

Office for Official Publications of the European Communities (1997) *Indicators of Sustainable Development: A Pilot Study following the Methodology of the United Nations Commission on Sustainable Development*, Office for Official Publications of the European Communities, Luxembourg.

Pankhurst, C, Doube, B & Gupta, V (eds) (1997) *Biological Indicators of Ecosystem Health*, CAB International, New York.

Philpott, L (1996) *CRC Project 1.3 Life Cycle Assessment*, CRC for Waste Management and Pollution Control Ltd.

Proops, JLR, Atkinson, G, Schlotheim, BF & Simon, S (1999) International trade and the sustainability footprint. *Ecological Economics*, 28, 75–9.

Rapport D, (1998a) Answering the critics. In D Rapport, R Costanza, PR Epstein, C Gaudet & R Levins (eds) *Ecosystem Health*, Blackwell Science, Malden USA.

Rapport, D (1998b) Defining ecosystem health. In D Rapport, R Costanza, PR Epstein, C Gaudet & R Levins (eds) *Ecosystem Health*, Blackwell Science, Malden USA.

SCARM (Standing Committee on Agriculture and Resource Management) (1998) *Sustainable Agriculture: Assessing Australia's Recent Performance*, CSIRO Publishing, Melbourne.

Shell (1998) *The Shell Report — 1998*. Web site: http://www.shell.com/download/2872/index.html.

South Sydney City Council (1995) *Strategy for a Sustainable City of South Sydney*, South Sydney City Council, Sydney.

Standards Australia (1996) *Environmental Management Systems: Specification with Guidance for Use*. Prepared by Joint Technical Committee QR/11, Environmental Management, Standards Australia, Sydney.

SEAC (State of the Environment Advisory Council) (ed. (1996) *Australia State of the Environment 1996*, CSIRO Publishing, Melbourne.

Thorman, R & Heath, I (1997) *Regional Environmental Strategies: How to Prepare and Implement Them*, Australian Local Government Association, Canberra.

UNCED (United Nations Conference on Environment and Development) (1992) *Agenda 21*, UNCED Secretariat, Geneva.

United Nations Division for Sustainable Development (1999) *Indicators of Sustainable Development*. Web sites: http://www.un.org/esa/sustdev/indisd/english/intro-duc.htm, http://www.un.org/esa/sustdev/indisd/english/worklist.htm.

UNEP (United Nations Environment Programme) (1996) *Engaging Stakeholders, Vols 1 and 2*, UNEP/SustainAbility Ltd, London.

Van Woerden, J (1999) *Data Issues of Global Environmental Reporting: Experiences from GEO-2000*, United Nations Environment Programme, Nairobi.

von Weizsacker, E, Lovins, AB & Lovins, LH (1997) *Factor Four Doubling Wealth — Halving Resource Use*, Allen and Unwin, Sydney.

Wackernagel, M & Rees, W (1996) *Our Ecological Footprint: Reducing Human Impact on the Earth*, New Society Publishers, Philadelphia.

Walker, J & Reuter, DJ (eds) (1996) *Indicators of Catchment Health*, CSIRO Publishing, Melbourne.

Wolman (1965) The metabolism of the city. *Scientific American*, 213, 79.

Woodley, S (1993) Monitoring and measuring ecosystem integrity in Canadian national parks. In S Woodley, J Kay & G Francis (eds) *Ecological Integrity and the Management of Ecosystems*, St Lucie Press, Delray Beach.

Yencken, D & Wilkinson, D (2000) *Resetting the Compass: Australia's Journey towards Sustainability*, CSIRO Publishing, Melbourne.

CHAPTER 4

Ayers, R & Kneese, A (1969) Production, consumption and externalities. *American Economic Review*, 59, 282–97.

Beckerman, W (1972) Economists, scientists and environmental catastrophe. *Oxford Economic Papers*, 24, 327–44.

Boulding, KE (1966) The economics of coming spaceship earth. In H Jarrett (ed.) *Environmental Quality in a Growing Economy*, Johns Hopkins Press, Baltimore.

Carson, R (1962) *Silent Spring*, Houghton Mifflin Co, Boston.

Common, M (1995) *Sustainability and Policy — Limits to Economics*, Cambridge University Press, Cambridge.

Daly, H (1992) Allocation, distribution and scale: Towards an economics that is efficient, just and sustainable. *Ecological Economics*, 6, 145–95.

Daly, H (1995) On Wilfred Beckerman's critique of sustainable development. *Environmental Values*, 4, 1.

Daly, H & Cobb, JB (1989) *For the Common Good*, Beacon Press, Boston.

De Steiguer, JE (1997) *The Age of Environmentalism*, McGraw Hill, New York.

Ehrlich, P (1968) *The Population Bomb*, Sierra Club and Ballantine Books, New York.

Georgescu-Roegen, N (1971) *The Entropy Law and the Economic Process*, Harvard University Press, Cambridge.

Gordon H Scott (1954) The economic theory of a common property resource, the fishery. *Journal of Political Economy*, 62, 124–42.

Hamilton, C (1994) *The Mystic Economist*, Willow Park Press, Australia.

Harcourt, GC (1982) The social science imperialists. In P Kerr (ed.) *The Social Science*

Imperialists: Selected Essays, Routledge and Kegan Paul, London and Boston.

Hardin, G (1968) The tragedy of the commons. *Science,* 162, 1243–8.

Hartwick, JM (1977) Intergenerational equity and the investing of rents from exhaustible resources. *American Economic Review,* 66, 972–74.

Hartwick, JM & Olewiler, ND (1998) *The Economics of Natural Resource Use,* Addison Wesley, (second edition).

Hatch, JH (1995) The market's hidden costs. *Australian Financial Review,* 14/11/95, p 19.

Hicks, J (1946) *Value and Capital,* Oxford University Press, London, (second edition).

Holland, A (1997) Sustainability. In J Foster (ed.) *Valuing Nature,* Routledge, London.

Jacobs, M (1991) *The Green Economy,* Pluto Press, London.

Krutilla, JV (1967) Conservation reconsidered. *American Economic Review,* 57, 777–86.

Kuhn, T (1970) *The Structure of Scientific Revolutions,* Chicago University Press, Chicago, (second edition).

Lancaster, K (1969) *Introduction to Modern Microeconomics,* Rand McNally and Company, Chicago.

Marshall, AJ (ed.) (1966) *The Great Extermination,* Heinemann, London.

Mill, JS (1848) *Principles of Political Economy with some of their Applications to Social Philosophy.* Ashley, WJ (ed.). Augustus M Kelley, New York, (1965 Edition).

Pearce, D (ed.) (1991) *Blueprint 2: Greening the World Economy,* Earthscan, London.

Pearce, D, Markandya, A & Barbier, EB (1989) *Blueprint for a Green Economy,* London: Earthscan.

Pearce, D & Turner, K (1990) *Economics of Natural Resources and the Environment,* Harvester Wheatsheaf, New York.

Pearce, D & Warford, JJ (1993) *World Without End,* Oxford University Press for the World Bank, New York.

Perring, M (1991) Ecological sustainability and environmental control. *Structural Change and Economic Dynamics,* 2, 275–95.

Singer, P (1975) *Animal Liberation,* Avon Books, New York.

Singer, P (1985) Prologue, ethics and the new animal liberation movement. In P Singer (ed.) *In Defence of Animals,* Blackwell, Oxford.

Smith, A (1776) *An Inquiry into the Nature and Causes of the Wealth of Nations,* Methuen, London, (1961 edition).

Solow, R (1974) Intergenerational equity and exhaustible resources. *Review of Economic Studies, Symposium,* 29–45.

Solow, R (1991) Sustainability: An Economist's Perspective. Eighteenth J Seward Johnson Lecture to the Marine Policy Centre, Woods Hole Oceanographic Institution, Mass.

Tietenberg, T (1994) *Environmental Economics and Policy,* Harper Collins, New York.

Turner R Kerry, Pearce, D & Bateman, I (1994) *Environmental Economics,* Harvester Wheatsheaf, New York.

Wills, I (1997) *Economics and the Environment,* Allen and Unwin, Sydney.

World Commission on Environment and Development (1987) *Our Common Future* (Brundtland Report), Oxford University Press, Oxford.

CHAPTER 5

Adriaanse, A (1993) *Environmental Policy Performance Indicators: A Study on the Development of Indicators for Environmental Policy for the Netherlands,* Sdu Uitgerverrij, Koninginnegracht, 175 pp.

ALGA (Australian Local Government Association) (1999) *Choosing and Using Environmental Indicators.* ALGA, Canberra, 20 pp.

ANZECC (2000) *Core Environmental Indicators for Reporting on the State of the*

Environment. ANZECC, Canberra, 92 pp.

ARMCANZ (Agriculture and Resource Management Council of Australia and New Zealand) (1998) *Sustainable Agriculture: Assessing Australia's Recent Performance*, SCARM Technical Report No. 70, CSIRO, Melbourne.

Australian Bureau of Statistics (2000) Website: http://www.abs.gov.au/websitedbs/ c311215.nsf/20564c23f3183fdaca25672100813ef1/36f96955a6e59068ca2568f 2001ae1cb/$FILE/Commonwealth+paper5.pdfInclude?

Australian National Audit Office (1996) *Performance Information Principles: Better Practice Guide*, Department of Finance, Canberra.

Commonwealth of Australia (1992) *National Strategy for Ecologically Sustainable Development*, Australian Government Publishing Service, Canberra.

DEHAA (Department for Environment, Heritage and Aboriginal Affairs) (1999) *Environmental Performance Measures: Signposts to the Future*. DEHAA, Adelaide.

DEST (Department of the Environment, Sport and Territories) (1994) *State of the Environment Reporting: Framework for Australia*, DEST, Canberra, 42 pp.

Environment Australia (2000) *Public Environmental Reporting: An Australian Approach*, Environment Australia, Canberra, 48 pp.

EPA (Environment Protection Authority) (1994*) Environment Protection (Waste Management) Policy 1994.* Department of Environment and Natural Resources, Adelaide.

EPA (Environment Protection Authority) (1997) *State of the Environment Reporting in South Australia: Position Paper*, Department of Environment and Natural Resources, Adelaide.

EPA (Environment Protection Authority) (1998) *State of the Environment Report for South Australia 1998.* Environment Protection Authority, Adelaide.

EPA NSW (Environment Protection Authority NSW) (1995a) *Environmental Guidelines. State of the Environment Reporting by Local Government*, EPA NSW, Sydney.

EPA NSW (Environment Protection Authority NSW) (1995b) *New South Wales State of the Environment 1995.* EPA, Sydney.

EPA NSW (Environment Protection Authority NSW) (1996) *The Future of NSW State of the Environment Reporting, Discussion Paper,* EPA, Sydney.

EPA NSW (Environment Protection Authority NSW) (1997a) *Corporate Environmental Reporting: How and Why,* EPA, Sydney.

EPA NSW (Environment Protection Authority NSW) (1997b) *New South Wales State of the Environment 1997,* EPA, Sydney.

EPC (Environment Protection Council of South Australia) (1988) *The State of the Environment Report for South Australia, 1988.* Department of Environment and Planning, Adelaide, 267 pp.

Government of Western Australia (1997) *Environment Western Australia. 1997 Draft State of the Environment Report.* Draft Report for Public Discussion. Department of Environmental Protection, Perth.

Government of Western Australia (1998) *Environment Western Australia. 1998 State of the Environment Report,* Department of Environmental Protection, Perth.

Government of Western Australia (1999) *Environmental Action: Government's Response to the State of the Environment Report,* Department of Environmental Protection, Perth.

Harding, R & Eckstein, D (1996) A Preliminary Discussion Paper on Development of Composite Indices for State of the Environment Reporting in NSW for Discussion at a 12 February Workshop. Summary. Prepared for the NSW EPA by the Institute of Environmental Studies, University of New South Wales. (Unpublished).

Ministry of Environment, Lands and Parks (1998) *Environmental Trends in British Columbia*, Ministry of Environment, Lands and Parks, British Columbia Web site: http://www.elp.gov.bc.ca/sppl/soerpt.

NEPC (National Environment Protection Council) (1998) *National Environment*

Protection (Ambient Air Quality) Measure, NEPC, Adelaide.

OECD (Organisation for Economic Co-operation and Development) (1994) *Environment Indicators. OECD Core Set*, OECD, Paris, 159 pp.

SABV (South Australian Business Vision 2010) (1999) *Making a difference through Benchmarking*, SABV 2010, Adelaide.

SABV (South Australian Business Vision 2010) (2000) *Making a difference through Benchmarking 2000*, SABV 2010, Adelaide. Website: http:/www.sabv2010.com.au

Saunders, D, Margules, C & Hill, B (1998) *Environmental Indicators for National State of the Environment Reporting: Biodiversity*, Environment Australia, Canberra.

SEAC (State of the Environment Advisory Council) (1996) *Australia State of the Environment 1996*, CSIRO, Melbourne.

SDAC (Sustainable Development Advisory Council) (1997) *State of the Environment Tasmania 1997 Volume 1 Condition and Trends*, SDAC, Hobart.

South Australian Department of Treasury and Finance (1998) *Budget Handbook*. Department of Treasury and Finance, Adelaide.

The Body Shop (1999) *The Body Shop Results 1998: The New Bottom Line*, The Body Shop, Melbourne.

Unilever (1998) *Environment Report 1998: Making Progress*, Unilever, London.

Venning, J (ed.) (1996) *Indicators for Better Environmental Management*. Proceedings of a seminar held in Adelaide on 9 February 1966. Department of Environment and Natural Resources, Adelaide, 79 pp.

Western Mining Corporation Limited (2000) *Environment Progress Report 1999*. WMC Ltd, Southbank.

CHAPTER 6

Cobb, C & Cobb, J (1994) *The Green National Product: A Proposed Index of Sustainable Economic Welfare*, University Press of America, Maryland.

Cobb, C, Halstead, T & Rowe, J (1995) *The Genuine Progress Indicator: Summary of Data and Methodology*, Redefining Progress, San Francisco.

Common, M (1995) *Sustainability and Policy — Limits to Economics*, Cambridge University Press, Cambridge.

Costanza, R, d'Arge, R, de Groot, R, Farber, S, Grasso, M, Hannon, B, Limburg, K, Naeem, S, O'Neill, R, Paruelo, J, Raskin, R, Sutton, P & van der Belt, M (1997) The value of the world's ecosystem services and natural capital. *Nature*. 387, 253–60.

Daly, H & Cobb, JB (1989) *For the Common Good: Redirecting the Economy toward Community, the Environment and a Sustainable Future*, Beacon Press, Boston.

Eisner, R (1985) *The Total Incomes Systems of Accounts*, Survey of Current Business, January.

Hamilton, C (1997) *The Genuine Progress Indicator: A New Index of Changes in Well-Being in Australia*, The Australia Institute, Canberra.

Heilbroner, R & Thurow, L (1998) *Economics Explained*, Touchstone (Simon and Schuster), New York.

Hicks, J (1948) *Value and Capital*, Oxford University Press, London, (second edition).

Nordhaus, W & Tobin, I (1972) Is growth obsolete? In NBER (National Bureau of Economic Research). *Economic Growth: Fifth Anniversary Colloquium*, NBER, New York.

CHAPTER 7

Adriaanse, A (1993) *Environmental Policy Performance Indicators: A study on the Development of Indicators for Environmental Policy for the Netherlands*, Sdu Uitgerverrij, Koninginnegracht, 175 pp.

Agricultural Council of Australia and New Zealand (1993) *Sustainable Agriculture:*

Tracking the indicators for Australia and New Zealand. Standing Committee on agriculture and resource management. Report no. 51. CSIRO, Melbourne.

ARMCANZ (Agriculture and Resource Management Council of Australia and New Zealand) (1998) *Sustainable Agriculture: Assessing Australia's Recent Performance,* SCARM Technical Report No. 70, CSIRO, Melbourne.

Australian Agricultural Council (1991) *Sustainable Agriculture.* Working Group on Sustainable Agriculture of Standing Committee on Agriculture, Technical Report No.36, CSIRO, Melbourne.

Australian Bureau of Statistics (1998, 1999) *Australian supplementary national accounts.*

Australian Bureau of Statistics (2000) *Proposed Headline Sustainability Indicators for Australia.* Web site: http://www.abs.gov.au/websitedbs/c311215.nsf/20564c23f3183fdaca2567210 0813ef1/36f96955a6e59068ca2568f2001ae1cb/$FILE/Commonwealth+paper 5.pdf.

Brown, LR (1995) *Can China Feed Itself?* Worldwatch Institute, Washington DC

Campbell, A (2000) Contrasting perspectives: young countries in old landscapes, and old landscapes in young countries. In A Hamblin (ed.) *Visions of Future Landscapes* Proceedings of the 1999 Fenner Conference on the Environment, 2–5 May 1999. Bureau of Rural Sciences, Canberra. (In press).

Canadian Forest Services (1995) *The Montreal Process: Criteria and Indicators for the Conservation and Sustainable Management of Temperate and Boreal Forests,* Canadian Forest Service, Quebec.

Cocks, D (1999) *Future Makers, Future Takers: Life in Australia in 2050,* CSIRO, Melbourne.

Commonwealth of Australia (1990) *Ecologically sustainable development: a Commonwealth discussion paper.* Australian Government Publishing Service, Canberra.

Commonwealth of Australia (1992a) *National Forest Policy Statement,* Australian Government Publishing Service, Canberra.

Commonwealth of Australia (1992b) *National Strategy for Ecologically Sustainable Development,* Australian Government Publishing Service, Canberra.

Commonwealth of Australia (1997) *The Natural Heritage Trust Act of Australia.*

Daly, H (1973) *Towards a Steady-state Economy,* WH Freeman, New York.

Daly, H (1988) On sustainable development and national accounts. In D Collard, D Pearce & D Ulph (eds) *Economics, Growth and Sustainable Environments,* MacMillan, London, pp 28–56.

Daly, H & Cobb, J (1989) *For the Common Good: Redirecting the Economy toward Community, the Environment and a Sustainable Future,* Beacon Press, Boston.

Eckersley, R (ed.) (1998) *Measuring Progress: Is Life Getting Better?* CSIRO Publishing, Melbourne, 382 pp.

Harding, R (ed.) (1996) *Tracking Progress: Linking Environment and Economy through Indicators and Accounting Systems.* Proceedings of the 1996 Fenner Conference on the Environment, 30 September – 3 October, University New South Wales, Sydney.

Garcia, SM, Staples, DJ & Chesson, J (1999) The FAO guidelines for the development and use of indicators for sustainable development of marine capture fisheries and an Australian example of their application. Paper CM 199/P:05 presented at the 1999 Science Conference of the International Council for the Exploration of the Sea, Stockholm.

Heilig, GK (1999) Can China feed itself? A system for Evaluation of Policy Options. CD-ROM. International Institute of Applied Systems Analysis, Austria. Web site: http://www.iiasa.ac.

IPCC (International Panel on Climate Change) (1995) *The Science of Climate Change: Summary for Policymakers.* Technical Report of IPCC Working Group 1.

IPCC/WMO/UNEP, UK Meteorological Office, Bracknell, 56 pp.

IPCC (International Panel on Climate Change) (1998) *IPCC Workshop on Rapid Non-linear Climate Change*. Noordwijkirhout, The Netherlands. IPCC UK Meteorological Office, Bracknell.

Lutz, E (ed.) (1993) *Towards Improved Accounting for the Environment*. An UNSTAT-World Bank Symposium. The World Bank, Washington.

Murray-Darling Basin Ministerial Council (1999) *The Salinity Audit of the Murray Darling Basin: A Hundred Year Perspective*, Murray-Darling Basin Commission, Canberra.

National Forest Inventory (1998) *Australia's State of the Forests Report*, Bureau of Rural Sciences, Canberra.

OECD (Organisation for Economic Co-operation and Development) (1998) *Towards Sustainable Development: Environmental Indicators*, OECD, Paris.

OECD (Organisation for Economic Co-operation and Development) (1999a) *Frameworks to Measure Sustainable Development*. An OECD Expert Workshop, Paris, 2–3 September 1999.

OECD (Organisation for Economic Co-operation and Development) (1999b) *Towards Sustainable Development Indicators to Measure Progress*. OECD Conference Proceedings, Rome, 15–17 December 1999.

Pearce, D & Anderson, G (1993) Capital theory and the measurement of sustainable development: An indicator of weak sustainability. *Ecological Economics*, 8, 103–8.

Productivity Commission (1999) *Implementation of Ecologically Sustainable Development by Commonwealth Departments and Agencies, Final Report*, Commonwealth of Australia, Canberra.

UNDP (United Nations Development Programme) (1990 to 1999 annually) *The Human Development Report*, Oxford University Press, New York.

Yencken, D & Wilkinson, D (2000) *Resetting the Compass: Australia's Journey towards Sustainability*, CSIRO Publishing, Melbourne.

Waring, M (1988) *Counting for nothing: What Men Value and what Women are Worth*. Allen and Unwin and Port Nicholson Press, New Zealand.

Whitworth, B, Chesson, J & Smith, T (2000) *Reporting on Ecologically Sustainable Development in Commonwealth Fisheries*. Draft Final Report to the Fisheries Resources Research Fund, July 2000, Bureau of Rural Sciences, Canberra.

World Bank (1999) *World Development Indicators 1999*. The World Bank, Washington DC

World Economic Forum (1999) *Pilot Environmental Sustainability Index*. Annual Meeting of the World Economic Forum, Davos, Switzerland, 2000 in collaboration with Yale Centre for Environmental Law and Policy, Yale University, and Centre for International Earth Science Information.

CHAPTER 8

Alcamo, J, Kreileman, GJJ, Krol, MS & Zuidema, G (1994) Modelling the global society-biosphere-climate system: part 1: Model description and testing. In *Image 2.0: Integrated Modelling of Global Climate Change*. Alcamo, J (ed.) Kluwer Academic Publishers, Dordrecht, The Netherlands.

Alcamo, J, Kreileman, E, Krol, M, Leemans, R, Bollen, J, van Minnen, J, Schaeffer, M, Toet, S & de Vries, B (1998) Global modelling of environmental change: An overview of IMAGE 2.1. In *Global Change Scenarios of the 21st Century: Results from the IMAGE 2.1 Model*. Alcamo, J, Leemans, R & Kreileman, E (eds) Pergamon, Elsevier Science, Oxford UK.

Ayers, RU (1998) *Turning Point: An End to the Growth Paradigm*. Earthscan Publications, London, 258 pp.

Brown, LR, Renner, M & Flavin, C (1998) *Vital Signs 1998: The Environmental Trends that are Shaping our Future*. Worldwatch Institute, Norton and Company, New York.

Cocks, D (1996) *People Policy: Australia's Population Choices.* UNSW Press, Sydney, 347 pp.

Cocks, D (1999) *Future Makers, Future Takers: Life in Australia 2050.* UNSW Press, Sydney, 332 pp.

Cohen, JE (1995) *How Many People Can the World Support?* Norton and Company, New York.

EPAC (Economic Planning Advisory Commission) (1994) *A Comparison of Economy-Wide Models of Australia: Responses to a Rise in Labour Productivity.* Papers resulting from an EMBA Symposium held in Canberra on 13 April 1994. Edited by Colin Hargreaves, October 1994.

Erlich, PR (1968) *The Population Bomb.* Ballantine Books, New York.

Forrester, JW (1961) *Industrial Dynamics.* The MIT Press, Cambridge Ma.

Foran, BD & Crane, D (1998) The OzECCO embodied energy model of Australia's physical economy. In *Advances in Energy Studies: Energy Flows in Ecology and Economy* Conference held at Porto Venere, Italy, 26–30 May 1998, pp 579–96.

Foran, BD & Mardon, C (1999) *Beyond 2025: Transitions to the Biomass-Alcohol Economy using Ethanol and Methanol.* CSIRO Resource Futures Working Document 99/07.

Gault, FD, Hamilton, KE, Hoffman, RB & McInnis, BC (1987) The design approach to socio-economic modelling. *Futures* 3–25, February.

Goudriaan, J, Schugart, HH, Bugmann, H, Cramer, W, Bondeau, A, Gardner, RH, Hunt, LA, Lauenroth, WK, Landsberg, JJ, Linder, S, Noble, IR, Parton, WJ, Pitelka, LF, Stafford, Smith, M, Sutherst, RW, Valentin, C & Woodward, FI (1999). Use of models in global change studies. In *The Terrestrial Biosphere and Global Change: Implications for Natural and Managed Ecosystems.* Walker, B, Steffen, W, Canadell, J and Ingram, J (eds) IGBP Book Series 4. Cambridge University Press, Cambridge UK.

Godet, M (1991) *From Anticipation to Action: A Handbook of Strategic Prospective.* UNESCO Publishing, Paris.

Long Term Strategies Committee (1994) *Australia's Population Carrying Capacity: One Nation-Two Ecologies.* House of Representatives Standing Committee for Long Term Strategies, Australian Government Publishing Service, Canberra.

Lenzen, M (1998). Primary energy and greenhouse gases embodied in Australian final consumption: An input-output analysis. *Energy Policy* 26, 6495–506.

Lutz, W (ed.) (1994) *Population-Development-Environment: Understanding their Interactions in Mauritius.* Springer-Verlag, Berlin.

Meadows, DH, Meadows, DL & Randers, J (1992) *Beyond The Limits: Global Collapse or a Sustainable Future.* Earthscan Publications, London.

Mentor Books (1960) Three Essays on Population: Thomas Malthus, Julian Huxley, Frederick Osborn, Mentor Books, New York.

McDonald, P & Kippen, R (1999) *The Impact of Immigration on the Ageing of Australia's Population.* Discussion Paper, Department of Immigration and Multicultural Affairs, May 1999, 36 pp.

Newman, P & Kenworthy, J (1999) *Sustainability and Cities: Overcoming Automobile Dependence.* Island Press, Washington DC

Robbert Associates (2000) Web Site. http://www.robbert.ca/. Accessed 20/04/2000.

Santa Fe Institute (2000) Web Site: http://www.santafe.edu/index.html.

Simon, JL (1990) *Population Matters: People, Resources, Environment and Immigration.* New Brunswick, NJ/London.

Slesser, M (1992) *ECCO User Manual Part 1. Third Edition.* The Resource Use Institute, Edinburgh, Scotland.

Slesser, M, King, J & Crane, DC (1997) *The Management of Greed: A BioPhysical Appraisal of Environmental and Economic Potential.* The Resource Use Institute, Edinburgh, Scotland, 332 pp.

Thomas, JF, Adams, P, Hall, N & Watson, B (1999) *Water and the Australian Economy.*

Australian Academy of Technological Sciences and Engineering, April 1999.

University of Berkeley (2000) *Sketch for a Historical Picture of the Progress of the Human Mind (1795).* Condorcet: The Education of Progress. Web site http://ishi.lib.berkeley.edu/~hist280/research/condorcet/pages/progress_main .html. Accessed 20 April 2000.

von Weizsacker, E, Lovins, AB & Lovins, LH (1997). *Factor 4: Doubling Wealth-Halving Resource Use.* Allen and Unwin, Sydney.

Waldrop, MM (1994) *Complexity: The Emerging Science at the Edge of Order and Chaos.* Penguin Books, London.

Wackernagel, M & Rees, W (1996) *Our Ecological Footprint: Reducing Human Impact on the Earth.* New Society Publishers, Gabriola Island and Philadelphia.

CHAPTER 9

ANAO (Australian National Audit Office) (1997) *Commonwealth Natural Resource Management and Environment Programs: Australia's Land, Water and Vegetation Resources,* Audit report 36, Australian Government Printing Service, Canberra.

Anderson, E & others (1997) Review of the national state of the environment report. *Australian Journal of Environmental Management,* 4, 157–84.

Aquatech Pty Ltd (1995) *Biodiversity Monitoring in Australia for Conservation and Ecologically Sustainable Use.* Unpublished report to the Biodiversity Unit, Commonwealth Department of the Environment, Sport and Territories, February 1995.

Boyden, S, Millar, S, Newcombe, K & O'Neill, B (1981) *The Ecology of a City and its People: The Case of Hong Kong,* Australian National University Press, Canberra.

Cardew, R & Fanning, P (1996) *Population-Environment-Economy Relationships in the Wollongong Area: A Framework for Analysis and Policy Development.* Australian Government Printing Service, Canberra.

Carpenter, SR, Chisholm, SW, Krebs, CJ, Schindler, DW & Wright, RF (1995) Ecosystem experiments. *Science,* 269, 324–27.

Common, MS (1995) *Sustainability and Policy: Limits to Economics,* Cambridge University Press, Melbourne.

Common, MS & Norton, TW (1994) Biodiversity, natural resource accounting and ecological monitoring. *Environmental and Resource Economics,* 4, 29–53.

Commonwealth of Australia (1992) *National Strategy for Ecologically Sustainable Development,* Australian Government Printing Service, Canberra.

Costanza, R, d'Arge, R, de Groot, R, Farber, S, Grasso, M, Hannon, B, Limburg, K, Naeem, S, O'Neill, R, Paruelo, J, Raskin, R, Sutton, P & van der Belt, M (1997) The value of the world's ecosystem services and natural capital. *Nature.* 387, 253–60.

Curtis, A, Robertson, A & Race, D (1998) Lessons from recent evaluations of natural resource management programs in Australia. *Australian Journal of Environmental Management,* 5, 109–19.

Dovers, S (1994) Historical and current patterns of energy use. In S Dovers (ed.) *Sustainable Energy Systems: Pathways to Australian Energy Reform,* Cambridge University Press, Melbourne.

Dovers, S (1995) Information: sustainability and policy. *Australian Journal of Environmental Management,* 2, 142–56.

Dovers, S (1997) Sustainability: demands on policy. *Journal of Public Policy,* 16, 303–18.

Dovers, S (1998) Dimensions of the Australian population-environment debate. *Development Bulletin,* 41, 50–3.

Dovers, S (1999) Institutionalising ecologically sustainable development. In K Walker & K Crowley (eds) *Australian Environmental Policy 2: Studies in Decline and Devolution,* UNSW Press, Sydney.

Dovers, S (2000a) *Beyond Everything Care and Everything Watch: Public Participation, Public Policy, and Participating Publics. Proceedings, International Landcare 2000 Conference.* Department of Natural Resources and Environment, Melbourne.

Dovers, S (ed.) (2000b) *Environmental History and Policy: Still Settling Australia,* Oxford University Press, Melbourne.

Dovers, S & Dore, J (1999) Adaptive institutions, organisations and policy processes for river basin and catchment management. In *Proceedings, 2nd International River Management Symposium, Brisbane, 29 September – 2 October,* Riverfestival, Brisbane, pp 123–33.

Dovers, S & Gullett, W (1999) Policy choice for sustainability: marketization, law and institutions. In K Bosselman & B Richardson (eds) *Environmental Justice and Market Mechanisms,* Kluwer Law International, London.

Dovers, S & Handmer, J (1999) Ignorance, sustainability, and the precautionary principle: towards an analytical framework. In R Harding & L Fisher (eds) *Perspectives on the Precautionary Principle,* Federation Press, Sydney.

Dovers, S & Mobbs, C (1997) An alluring prospect? Ecology, and the requirements of adaptive management. In N Klomp & I Lunt (eds) *Frontiers in Ecology: Building the Links,* Elsevier, London.

Dovers, S, Norton, T & Handmer, J (1996) Uncertainty, ecology, sustainability and policy. *Biodiversity and Conservation,* 5, 1143–167.

Dovers, S, Norton, T, Hughes, I & Day, L (1992) *Population growth and Australian regional environments.* Australian Government Publishing Service, Canberra.

Eckerlsey, R (ed.) (1995) *Markets, the State and the Environment: Towards Integration,* Macmillan, Melbourne.

Goodin, RE (1995) Institutions and their design. In RE Goodin (ed.) *The Theory of Institutional Design,* Cambridge University Press, Cambridge.

Gunderson, L, Holling, C & Light, S (eds) (1995) *Barriers and Bridges to the Renewal of Ecosystems and Institutions,* Columbia University Press, New York.

Hamilton, C & Throsby, D (eds) (1998) *The Ecologically Sustainable Development Process: Evaluating a Policy Experiment,* Academy of the Social Sciences, Canberra.

Henningham, R (ed.) (1995) *Institutions in Australian Society,* Oxford University Press, Melbourne.

Industry Commission (1998) *A Full Repairing Lease: Inquiry into Ecologically Sustainable Land Management.* Australian Government Printing Service, Canberra.

Just, J, Cardew, R, Andrews, A, Fairweather, P, Lawton, F & McLean, N (1995) Creating an EIA database. *Australian Journal of Environmental Management,* 2, 40–51.

Lindenmayer, DB (2000) Factors at multiple scales affecting distribution patterns and their implications for animal conservation — Leadbeater's Possum as a case study. *Biodiversity and Conservation,* 9, 15–35.

Lindenmayer, DB, Cunningham, RB, Pope, ML & Donelly, CF (1999) The response of arboreal marsupials to landscape context: a large-scale fragmentation study. *Ecological Applications,* 9, 594–611.

Lovett, S (1997) *Revitalising Rural Research and Development in Australia: The Story So Far,* Land and Water Resources R&D Corporation, Canberra.

May, P (1992) Policy learning and policy failure. *Journal of Public Policy,* 12, 331–54.

Nordhaus, WD & Kokkelenberg, EC (eds) (1999) *Nature's Numbers: Expanding the National Economic Accounts to include the Environment,* National Academy Press, Washington DC.

Norton, T, Dovers, S, Nix, H & Elias, D (1994) *An Overview of Research on the Links between Human Population and the Environment,* Australian Government Printing Service, Canberra.

Peters, R (1991) *A Critique for Ecology,* Cambridge University Press, Cambridge.

Productivity Commission (1999) *Implementation of Ecologically Sustainable Development by Commonwealth Departments and Agencies.* Ausinfo, Canberra.

Redhead, T, Mummery, J & Kenchington, R (1993) *Options for a National Program on long-term Monitoring of Australian Biodiversity*. (Unpublished).

Schrader-Frechette, K (1995) Hard ecology, soft ecology, and ecosystem integrity. In L Westra & J Lemons (eds) *Perspectives on Ecosystem Integrity*, Kluwer, Dordrecht.

Simpson, R, Petroeschevsky, A & Lowe, I (2000) An ecological footprint analysis for Australia. *Australian Journal of Environmental Management*, 7, 11–8.

SMEC/AIG (Snowy Mountains Engineering Corporation/Australian Industry Group) (2000) *A Framework for Public Environmental Reporting — An Australian Approach*, Environment Australia, Canberra.

Smithson, M (1989) *Ignorance and Uncertainty: Emerging Paradigms*, Springer-Verlag, New York.

Stein Justice P (2000) Are decision-makers too cautious with the precautionary principle? *Environmental and Planning Law Journal*, 17: 3–23.

SEAC (State of Environment Advisory Council) (1996) Adaptive management, *see* continuous improvement

: *State of the Environment 1996*, CSIRO Publishing, Melbourne.

Stewart, A (1999) Scientific uncertainty, ecologically sustainable development and the precautionary principle. *Griffith Law Review*, 8, 350–73.

Stewart, D & McColl, G (1994) The Resource Assessment Commission: and inside assessment. *Australian Journal of Environmental Management*, 1, 12–23.

Toyne, P (1994) *The Reluctant Nation*, ABC Books, Sydney.

United Nations (1992) *Agenda 21: The Programme of Action from Rio*, Oxford University Press, Oxford.

UNDP (United Nations Development Programme) (2000) *Human Development Report 2000*, Oxford University Press, Oxford.

Walker, K (1994) *The Political Economy of Environmental Policy: An Australian Introduction*, UNSW Press, Sydney.

WCED (World Commission on Environment and Development) (1987) *Our Common Future*, Oxford University Press, Oxford.

Wilson, EO (1993) *The Diversity of Life*, Penguin, London.

Wynne, B (1992) Uncertainty and environmental learning: reconceiving science in the preventative paradigm. *Global Environmental Change*, 2, 111–27.

Yencken, D & Wilkinson, D (2000) *Resetting the Compass: Australia's Journey towards Sustainability*, CSIRO Publishing, Melbourne.

INDEX